THE SCIENCE QUESTION IN FEMINISM

The Science Question
in Feminism

SANDRA HARDING

Cornell University Press

ITHACA AND LONDON

First published 1986 by Cornell University Press.
First printing, Cornell Paperbacks, 1986.
Second printing, 1986.

Printed in the United States of America

*The paper in this book is acid-free and meets the guidelines for
permanence and durability of the Committee on Production Guidelines
for Book Longevity of the Council on Library Resources.*

Library of Congress Cataloging-in-Publication Data

Harding, Sandra G.
    The science question in feminism.

    Bibliography: p.
    Includes index.
    1. Women in science. 2. Feminism. 3. Science—
Social aspects. 4. Sexism. I. Title.
HQ1397.H28 1986   305.4'2   85–48197
ISBN 0-8014-1880-1
ISBN 0-8014-9363-3 (pbk.)

CONTENTS

Contents

ACKNOWLEDGMENTS

This book could not have been written without the inspiration and support of many individuals and institutions. I am especially indebted to

. . . the women scientists, past and present, whose courageous struggles and hard-won achievements have made it possible for women today to gain scientific educations—and enabled us all to begin to understand why that can be only the beginning of the feminist revolution in science.

. . . the Invisible College of feminist theorists and science critics, whose brilliant and risk-taking work has provided me with exemplary models of feminist theory. For invaluable comments on my papers, parts of earlier versions of this book, and my thinking, I am particularly grateful to Margaret Andersen, Elizabeth Fee, Jane Flax, Donna Haraway, and Nancy Hartsock.

. . . Margaret Andersen and Nancy Hartsock for warm and loving collaboration in political and intellectual projects through the years this book was emerging.

. . . Frank Dilley for his continual efforts to ensure that departmental governance treats women fairly, too.

. . . The National Endowment for the Humanities, the National Science Foundation, and the University of Delaware for summer grants permitting me to work up some of the issues; the Fund for the Improvement of Post-Secondary Education for a Mina Shaughnessy scholarship; the Mellon Foundation for a fellowship at the Wellesley Center for Research on Women; and the University of Delaware for

7

Acknowledgments

a sabbatical—all of which provided the time necessary to bring this work to completion.

... Mary Imperatore for her willing, cheerful, and unerring secretarial support both before and after the arrival of Kaypro.

... My daughters, Dorian and Emily, for their loving enthusiasm for my ventures and for the inspiration provided by the courage, brilliance, and wit with which they approach their own.

SANDRA HARDING

Newark, Delaware

PREFACE

Since the mid-1970s, feminist criticisms of science have evolved from a reformist to a revolutionary position, from analyses that offered the possibility of improving the science ,we have, to calls for a transformation in the very foundations both of science and of the cultures that accord it value. We began by asking, "What is to be done about the situation of women in science?"—the "woman question" in science. Now feminists often pose a different question: "Is it possible to use for emancipatory ends sciences that are apparently so intimately involved in Western, bourgeois, and masculine projects?"—the "science question" in feminism.

The radical feminist position holds that the epistemologies, metaphysics, ethics, and politics of the dominant forms of science are androcentric and mutually supportive; that despite the deeply ingrained Western cultural belief in science's intrinsic progressiveness, science today serves primarily regressive social tendencies; and that the social structure of science, many of its applications and technologies, its modes of defining research problems and designing experiments, its ways of constructing and conferring meanings are not only sexist but also racist, classist, and culturally coercive. In their analyses of how gender symbolism, the social division of labor by gender, and the construction of individual gender identity have affected the history and philosophy of science, feminist thinkers have challenged the intellectual and social orders at their very foundations.

These feminist critiques, which debunk much of what we value in modern Western culture, appear to emerge from outside this culture.

9

That is indeed the case insofar as women have been excluded from the processes of defining the culture and have been conceived as the "other" against which men in power define their projects. Yet such destabilizing, "exploding," of the categories of social practice and thought is firmly within the tradition of modern Western history and its explicit commitment to criticism of traditional social practices and beliefs. One such belief is that androcentrism is "natural" and right; another is faith in the progressiveness of scientific rationality. From this perspective, the feminist critiques of science may be seen as calling for a more radical intellectual, moral, social, and political revolution than the founders of modern Western cultures could have imagined. Historically, it is just such revolutions—and not the process of scientific inquiry alone—that have fostered the development of progressive kinds of knowledge-seeking.

This book examines important trends in the feminist critiques of science with the aim of identifying tensions and conflicts between them, inadequate concepts informing their analyses, unrecognized obstacles to and gaps in their research programs, and extensions that might transform them into even more powerful tools for the construction of emancipatory meanings and practices. Motivating my investigation is the belief that these feminist science critiques can be shown to have implications at least as revolutionary for modern Western cultural self-images as feminist critiques in the humanities and social sciences have had.

It should not need to be said—but probably does—that I do not wish to be understood as recommending that we throw out the baby with the bathwater. We do not imagine giving up speaking or writing just because our language is deeply androcentric; nor do we propose an end to theorizing about social life once we realize that thoroughly androcentric perspectives inform even our feminist revisions of the social theories we inherit. Similarly, I am not proposing that humankind would benefit from renouncing attempts to describe, explain, and understand the regularities, underlying causal tendencies, and meanings of the natural and social worlds just because the sciences we have are androcentric. I am seeking an end to androcentrism, not to systematic inquiry. But an end to androcentrism will require far-reaching transformations in the cultural meanings and practices of that inquiry.

The first two chapters provide an overview and theoretical introduction. Chapter 1 identifies five feminist critiques and three feminist epistemological programs, and points to the challenges each of these faces. Chapter 2 looks at some problems in the understanding of both science and gender in the feminist science criticisms, and shows how these create obstacles to the development of a feminist theory of science; I then develop the more adequate concepts of science and gender that inform the following chapters.

The next three chapters show the connections between the parts of the picture of science that feminist critics have produced, and identify inconsistencies and oversights. Chapter 3 reviews the feminist approaches to equity issues in the structure of science and points to the tensions between these ahistorical images and the reality of science's social structure. Chapter 4 scrutinizes the feminist charges of androcentrism in the selection of problematics (of what is defined as requiring ·scientific explanation) and the design of research in biology and the social sciences (I include the social sciences here to prepare for later analysis of the inadequate social assumptions that have guided the mainstream understandings of modern science). Chapter 5 examines science's contribution to the construction of gendered meanings for both nature and inquiry and reviews the literature showing that much of what is commonly taken to be biological sex difference and sexual desire is socially constructed.

Chapters 6 and 7 turn to feminist theories of knowledge, the epistemological grounds for modern science, and the alternative justificatory strategies proposed by feminist critics. Chapter 6 examines the "successor science" projects of four theorists—Hilary Rose, Dorothy Smith, Jane Flax, and Nancy Hartsock—and their attempts to envision forms and purposes of knowledge-seeking that are alternative to those used to justify the science we have. In Chapter 7 I describe some obstacles that these epistemologies face; by focusing on the relationship between these feminist projects and similar emancipatory science projects of ex-colonial peoples, I also consider some of the difficult questions the "successor science" projects and feminist postmodernist critiques pose for each other.

Chapters 8 and 9 provide a pause in the argument by returning to the history of science in an effort to account for the deterioration of socially progressive knowledge-seeking (readers who prefer plots uninterrupted by the ghostly appearances of the protagonist's garrulous ancestors may want to skip to Chapter 10). Chapter 8, which treats the institution of science as a personage passing from infancy to adulthood, identifies gaps in the standard stories this adult personage tells about its infancy. Chapter 9 examines one kind of attempt by recent social histories of science to fill these gaps, and argues that even they tend to repress what they need to redress by systematically avoiding consideration of gender symbolism and actual social relations between the genders in history.

Chapter 10 returns to the main plot to reflect on some central instabilities and tensions within the feminist theories I have been examining and developing. It identifies questions asked by the science critiques that cannot be answered in the terms in which they have been posed. I conclude by pointing to the way feminist science critiques have assumed a reversal of the "unity of science" thesis so central to the members of the Vienna Circle. For feminists, it is moral and political, rather than scientific, discussion that has served as the paradigm—though a problematic one—of rational discourse.

S.H.

THE SCIENCE QUESTION IN FEMINISM

1 FROM THE WOMAN QUESTION IN SCIENCE TO THE SCIENCE QUESTION IN FEMINISM

Feminist scholars have studied women, men, and social relations between the genders within, across, and insistently against the conceptual frameworks of the disciplines. In each area we have come to understand that what we took to be humanly inclusive problematics, concepts, theories, objective methodologies, and transcendental truths are in fact far less than that. Instead, these products of thought bear the mark of their collective and individual creators, and the creators in turn have been distinctively marked as to gender, class, race, and culture.[1] We can now discern the effects of these cultural markings in the discrepancies between the methods of knowing and the interpretations of the world provided by the creators of modern Western culture and those characteristic of the rest of us. Western culture's favored beliefs mirror in sometimes clear and sometimes distorting ways not

[1] I make a sharp distinction between "sex" and "gender" (even though this is a dichotomy I shall later problematize); thus I refer to "gender roles" rather than "sex roles," etc., retaining only a few terms such as "sexism," where the substitution seems more distracting than useful. Otherwise (except in direct quotations), I use "sex" only when it is, indeed, biology that is at issue. There are two reasons for this policy. First, in spite of feminist insistence for decades, perhaps centuries, that women's and men's "natures" and activities are primarily shaped by social relations, not by immutable biological determinants, many people still do not grasp this point or are unwilling to commit themselves to its full implications (the current fascination with sociobiology is just one evidence of this problem). Second, the very thought of sex exerts its own fatal attraction for many otherwise well-intentioned people: such phrases as "sexual politics," "the battle between the sexes," and "male chauvinism" make the continuation of gender hostilities sound far more exciting than feminism should desire.

the world as it is or as we might want it to be, but the social projects of their historically identifiable creators.

The natural sciences are a comparatively recent subject of feminist scrutiny. The critiques excite immense anticipation—or fear—yet they remain far more fragmented and less clearly conceptualized than feminist analyses in other disciplines.

The anticipation and fear are based in the recognition that we are a scientific culture, that scientific rationality has permeated not only the modes of thinking and acting of our public institutions but even the ways we think about the most intimate details of our private lives. Widely read manuals and magazine articles on child rearing and sexual relations gain their authority and popularity by appealing to science. And during the last century, the social use of science has shifted: formerly an occasional assistant, it has become the direct generator of economic, political, and social accumulation and control. Now we can see that the hope to "dominate nature" for the betterment of the species has become the effort to gain unequal access to nature's resources for purposes of social domination. No longer is the scientist—if he ever was—an eccentric and socially marginal genius spending private funds and often private time on whatever purely intellectual pursuits happen to interest him. Only very rarely does his research have no foreseeable social uses. Instead, he (or, more recently, she) is part of a vast work force, is trained from elementary school on to enter academic, industrial, and governmental laboratories where 99 + percent of the research is expected to be immediately applicable to social projects. If these vast industrialized empires, devoted—whether intentionally or not—to material accumulation and social control, cannot be shown to serve the best interests of social progress by appeal to objective, dispassionate, impartial, rational knowledge-seeking, then in our culture they cannot be legitimated at all. Neither God nor tradition is privileged with the same credibility as scientific rationality in modern cultures.

Of course, feminists are not the first group to scrutinize modern science in this way. Struggles against racism, colonialism, capitalism, and homophobia, as well as the counter culture movement of the 1960s and the contemporary ecology and antimilitarism movements, have all produced pointed analyses of the uses and abuses of science. But the feminist criticisms appear to touch especially raw nerves. For one thing, at their best they incorporate the key insights of these other movements while challenging the low priority that specifically feminist concerns have been assigned in such agendas for social reform. For

16

another, they question the division of labor by gender—a social aspect of the organization of human relations that has been deeply obscured by our perceptions of what is "natural" and what is social. Perhaps most disturbingly, they challenge our sense of personal identity at its most prerational level, at the core. They challenge the desirability of the gendered aspects of our personalities and the expression of gender in social practices, which for most men and women have provided deeply satisfying parts of self-identity.

Finally, as a symbol system, gender difference is the most ancient, most universal, and most powerful origin of many morally valued conceptualizations of everything else in the world around us. Cultures assign a gender to such nonhuman entities as hurricanes and mountains, ships and nations. As far back in history as we can see, we have organized our social and natural worlds in terms of gender meanings within which historically specific racial, class, and cultural institutions and meanings have been constructed. Once we begin to theorize gender—to define gender as an analytic category within which humans think about and organize their social activity rather than as a natural consequence of sex difference, or even merely as a social variable assigned to individual people in different ways from culture to culture—we can begin to appreciate the extent to which gender meanings have suffused our belief systems, institutions, and even such apparently gender-free phenomena as our architecture and urban planning. When feminist thinking about science is adequately theorized, we will have a clearer grasp of how scientific activity is and is not gendered in this sense.

Now it is certainly true that racism, classism, and cultural imperialism often more deeply restrict the life opportunities of individuals than does sexism. We can easily see this if we compare the different life opportunities available to women of the same race but in different classes, or of the same class but in different races, in the United States today or at any other time and place in history. Consequently, it is understandable why working-class people and victims of racism and imperialism often place feminist projects low on their political agendas. Furthermore, gender appears only in culturally specific forms. As we shall see in the next chapter, gendered social life is produced through three distinct processes: it is the result of assigning dualistic gender metaphors to various perceived dichotomies that rarely have anything to do with sex differences; it is the consequence of appealing to these gender dualisms to organize social activity, of dividing necessary social activities between different groups of humans; it is a form of socially constructed individual identity only imperfectly correlated with either the "reality" or the perception of sex differences. I shall be referring to these three aspects of gender as *gender symbolism* (or, borrowing a term from anthropology, "gender totemism"), *gender structure* (or the division of labor by gender), and *individual gender*. The referents for all three meanings of masculinity and femininity differ from culture to culture, though within any culture the three forms of gender are related to each other. Probably few, if any, symbolic, institutional, or individual identity or behavioral expressions of masculinity and femininity can be observed in all cultures or at all times in history.

But the fact that there are class, race, and cultural differences between women and between men is not, as some have thought, a reason to find gender difference either theoretically unimportant or politically irrelevant. In virtually every culture, gender difference is a pivotal way in which humans identify themselves as persons, organize social relations, and symbolize meaningful natural and social events and processes. And in virtually all cultures, whatever is thought of as manly is more highly valued than what is thought of as womanly. Moreover, we need to recognize that in cultures stratified by both gender and race, gender is always also a racial category and race a gender category. That is, sexist public policies are different for people of the same gender but different race, and racist policies are different for women and men within the same race. One commentator has proposed that we think of these policies as, respectively, racist sexism and sexist racism.[2]

Finally, we shall later examine the important role to be played in emancipatory epistemologies and politics by open recognition of gender differences within racial groups and racial and cultural differences within gender groups. "Difference" can be a slippery and dangerous rallying point for inquiry projects and for politics, but each emancipatory struggle needs to recognize the agendas of other struggles as integral parts of its own in order to succeed. (After all, people of color come in at least two genders, and women are of many colors.) For each struggle, epistemologies and politics grounded in solidarities could replace the problematic ones that appeal to essentialized identities, which are, perhaps, spurious.

[2]Boch (1983). See also Caulfield (1974); Davis (1971). (Works cited in my notes by author and year of publication receive full citation in the bibliography, which lists the sources I have found most useful for this study. Additional references appear in full in the footnotes.)

For all these reasons, feminist critiques claiming that science, too, is gendered appear deeply threatening to the social order, even in societies such as ours where racism, classism, and imperialism also direct all our lives. Obviously, the different forms of domination use one another as resources and support one another in complex ways. If we find it difficult to imagine the day-to-day details of living in a world no longer structured by racism and classism, most of us do not even know how to start imagining a world in which gender difference, in its equation of masculinity with authority and value, no longer constrains the ways we think, feel, and act. And the day-to-day world we live in is so permeated by scientific rationality as well as gender that to nonfeminists and perhaps even some feminists, the very idea of a feminist critique of scientific rationality appears closer to blasphemy than to social-criticism-as-usual.

Feminists in other fields of inquiry have begun to formulate clear and coherent challenges to the conceptual frameworks of their disciplines. By putting women's perspective on gender symbolism, gender structure, and individual gender at the center of their thinking, they have been able to reconceive the purposes of research programs in anthropology, history, literary criticism, and so forth.[3] They have begun to retheorize the proper subject matters of the understandings these disciplines could provide. But I think the proper subject matters and purposes of a feminist critique of science have, thus far, eluded the firm grip and the clear conceptualizations that are becoming evident in much of this other research. The voice of feminist science criticism alternates among five different kinds of projects, each with its own audience, subject matter, ideas of what science is and what gender is, and set of remedies for androcentrism. In certain respects, the assumptions guiding these analyses directly conflict. It is not at all clear how their authors conceive of the theoretical connections between them, nor, therefore, what a comprehensive strategy for eliminating androcentrism from science would look like. This is particularly troublesome because clarity about so fundamental a component of our culture can have powerful effects elsewhere in feminist struggles.

One problem may be that we have been so preoccupied with responding to the sins of contemporary science in the same terms our culture uses to justify these sins that we have not yet given adequate attention to envisioning truly emancipatory knowledge-seeking. We

[3]McIntosh (1983).

19

have not yet found the space to step back and image up the whole picture of what science might be in the future. In our culture, reflecting on an appropriate model of rationality may well seem a luxury for the few, but it is a project with immense potential consequences: it could produce a politics of knowledge-seeking that would show us the conditions necessary to transfer control from the "haves" to the "have-nots."

What kind of understanding of science would we have if we began not with the categories we now use to grasp its inequities, misuses, falsities, and obscurities but with those of the biologist protagonist imagined by Marge Piercy in *Woman on the Edge of Time*, who can shift her/his sex at will and who lives in a culture that does not institutionalize (i.e., does not have) gender? or with the assumptions of a world where such categories as machine, human, and animal are no longer either distinct or of cultural interest, as in Anne McCaffrey's *The Ship Who Sang*?[4] Perhaps we should turn to our novelists and poets for a better intuitive grasp of the theory we need. Though often leaders in the political struggles for a more just and caring culture, they are professionally less conditioned than we to respond point by point to a culture's defenses of its ways of being in the world.

FIVE RESEARCH PROGRAMS

To draw attention to the lack of a developed feminist theory for the critique of the natural sciences is not to overlook the contributions these young but flourishing lines of inquiry have made. In a very short period of time, we have derived a far clearer picture of the extent to which science, too, is gendered. Now we can begin to understand the economic, political, and psychological mechanisms that keep science sexist and that must be eliminated if the nature, uses, and valuations of knowledge-seeking are to become humanly inclusive ones. Each of these lines of inquiry raises intriguing political and conceptual issues, not only for the practices of science and the ways these practices are legitimated but also for each other. Details of these research programs are discussed in following chapters; I emphasize here the problems they raise primarily to indicate the undertheorization of the whole field.

[4]Marge Piercy, *Woman on the Edge of Time* (New York: Fawcett, 1981); Anne McCaffrey, *The Ship Who Sang* (New York: Ballantine, 1976). Donna Haraway (1985) discusses the potentialities that McCaffrey's kind of antidualism opens up for feminist theorizing.

20

First of all, equity studies have documented the massive historical resistance to women's getting the education, credentials, and jobs available to similarly talented men;[5] they have also identified the psychological and social mechanisms through which discrimination is informally maintained even when the formal barriers have been eliminated. Motivation studies have shown why boys and men more often want to excel at science, engineering, and math than do girls and women.[6] But should women want to become "just like men" in science, as many of these studies assume? That is, should feminism set such a low goal as mere equality with men? And to which men in science should women want to be equal—to underpaid and exploited lab technicians as well as Nobel Prize winners? Moreover, should women want to contribute to scientific projects that have sexist, racist, and classist problematics and outcomes? Should they want to be military researchers? Furthermore, what has been the effect of women's naiveté about the depth and extent of masculine resistance—that is, would women have struggled to enter science if they had understood how little equity would be produced by eliminating the formal barriers against women's participation?[7] Finally, does the increased presence of women in science have any effect at all on the nature of scientific problematics and outcomes?

Second, studies of the uses and abuses of biology, the social sciences, and their technologies have revealed the ways science is used in the service of sexist, racist, homophobic, and classist social projects. Oppressive reproductive policies; white men's management of all women's domestic labor; the stigmatization of, discrimination against, and medical "cure" of homosexuals; gender discrimination in workplaces—all these have been justified on the basis of sexist research and maintained through technologies, developed out of this research, that move control of women's lives from women to men of the dominant group.[8] Despite the importance of these studies, critics of the sexist uses of science often make two problematic assumptions: that there is a value-free, pure scientific research which can be distinguished from the social uses of science; and that there are proper uses of science with which we

<hr>

[5] See, e.g., Rossiter (1982b); Walsh (1977).
[6] See Aldrich (1978).
[7] Rossiter (1982b) makes this point.
[8] See Tobach and Rosoff (1978; 1979; 1981; 1984); Brighton Women and Science Group (1980); Ehrenreich and English (1979); Rothschild (1983); Zimmerman (1983); Arditti, Duelli-Klein, and Minden (1984).

can contrast its improper uses. Can we really make these distinctions? Is it possible to isolate a value-neutral core from the uses of science and its technologies? And what distinguishes improper from proper uses? Furthermore, each misuse and abuse has been racist and classist as well as oppressive to women. This becomes clear when we note that there are different reproductive policies, forms of domestic labor, and forms of workplace discrimination mandated for women of different classes and races even within U.S. culture at any single moment in history. (Think, for instance, of the current attempt to restrict the availability of abortion and contraceptive information for some social groups at the same time that sterilization is forced on others. Think of the resuscitation of scientifically supported sentimental images of motherhood and nuclear forms of family life for some at the same time that social supports for mothers and nonnuclear families are systematically withdrawn for others.) Must not feminism take on as a central project of its own the struggle to eliminate class society and racism, homophobia and imperialism, in order to eliminate the sexist uses of science?

Third, in the critiques of biology and the social sciences, two kinds of challenges have been raised not just to the actual but to the possible existence of any pure science at all.[9] The selection and definition of problematics—deciding what phenomena in the world need explanation, and defining what is problematic about them—have clearly been skewed toward men's perception of what they find puzzling. Surely it is "bad science" to assume that men's problems are everyone's problems, thereby leaving unexplained many things that women find problematic, and to assume that men's explanations of what they find problematic are undistorted by their gender needs and desires. But is this merely—or, perhaps, even—an example of bad science? Will not the selection and definition of problems always bear the social fingerprints of the dominant groups in a culture? With these questions we glimpse the fundamental value-ladenness of knowledge-seeking and thus the impossibility of distinguishing between bad science and science-as-usual. Furthermore, the design and interpretation of research again and again has proceeded in masculine-biased ways. But if problems are necessarily value-laden, if theories are constructed to explain

---

[9]The literature here is immense. For examples of these criticisms, see Longino and Doell (1983); Hubbard, Henifin, and Fried (1982); Gross and Averill (1983); Tobach and Rosoff (1978; 1979; 1981; 1984); Millman and Kanter (1975); Andersen (1983); Westkott (1979).

problems, if methodologies are always theory-laden, and if observations are methodology-laden, can there be value-neutral design and interpretation of research? This line of reasoning leads us to ask whether it is possible that some kinds of value-laden research are nevertheless maximally objective. For example, are overtly antisexist research designs inherently more objective than overtly sexist or, more important, "sex-blind" (i.e., gender-blind) ones? And are antisexist inquiries that are also self-consciously antiracist more objective than those that are not? There are precedents in the history of science for preferring the distinction between objectivity-increasing and objectivity-decreasing social values to the distinction between value-free and value-laden research. A different problem is raised by asking what implications these criticisms of biology and social science have for areas such as physics and chemistry, where the subject matter purportedly is physical nature rather than social beings ("purportedly" because, as we shall see, we must be skeptical about being able to make any clear distinctions between the physical and the nonphysical). What implications could these findings and this kind of reasoning about objectivity have for our understanding of the scientific world view more generally?

Fourth, the related techniques of literary criticism, historical interpretation, and psychoanalysis have been used to "read science as a text" in order to reveal the social meanings—the hidden symbolic and structural agendas—of purportedly value-neutral claims and practices.[10] In textual criticism, metaphors of gender politics in the writings of the fathers of modern science, as well as in the claims made by the defenders of the scientific world view today, are no longer read as individual idiosyncrasies or as irrelevant to the meanings science has for its enthusiasts. Furthermore, the concern to define and maintain a series of rigid dichotomies in science and epistemology no longer appears to be a reflection of the progressive character of scientific inquiry; rather, it is inextricably connected with specifically masculine—and perhaps uniquely Western and bourgeois—needs and desires. Objectivity vs. subjectivity, the scientist as knowing subject vs. the objects of his inquiry, reason vs. the emotions, mind vs. body—in each case the former has been associated with masculinity and the latter with femininity. In each case it has been claimed that human progress requires the former to achieve domination of the latter.[11]

[10]Good examples are Keller (1984); Merchant (1980); Griffin (1978); Flax (1983); Jordanova (1980); Bloch and Bloch (1980); Harding (1980).

[11]The key "object-relations" theorists among these textual critics are Dinnerstein (1976); Chodorow (1978); Flax (1983). See also Balbus (1982).

Valuable as these textual criticisms have been, they raise many questions. What relevance do the writings of the fathers of modern science have to contemporary scientific practice? What theory would justify regarding these metaphors as fundamental components of scientific explanations? How can metaphors of gender politics continue to shape the cognitive form and content of scientific theories and practices even when they are no longer overtly expressed? And can we imagine what a scientific mode of knowledge-seeking would look like that was not concerned to distinguish between objectivity and subjectivity, reason and the emotions?

Fifth, a series of epistemological inquiries has laid the basis for an alternative understanding of how beliefs are grounded in social experiences, and of what kind of experience should ground the beliefs we honor as knowledge.[12] These feminist epistemologies imply a relation between knowing and being, between epistemology and metaphysics, that is an alternative to the dominant epistemologies developed to justify science's modes of knowledge-seeking and ways of being in the world. It is the conflicts between these epistemologies that generate the major themes of this study.

A GUIDE TO FEMINIST EPISTEMOLOGIES

The epistemological problem for feminism is to explain an apparently paradoxical situation. Feminism is a political movement for social change. But many claims, clearly motivated by feminist concerns, made by researchers and theorists in the social sciences, in biology, and in the social studies of the natural sciences appear more plausible— more likely to be confirmed by evidence—than the beliefs they would replace. How can such politicized research be increasing the objectivity of inquiry? On what grounds should these feminist claims be justified?

We can usefully divide the main feminist responses to this apparent paradox into two relatively well-developed solutions and one agenda for a solution. I will refer to these three responses as *feminist empiricism*, the *feminist standpoint*, and *feminist postmodernism*.

Feminist empiricism argues that sexism and androcentrism are social biases correctable by stricter adherence to the existing methodological norms of scientific inquiry. Movements for social liberation "make it

[12]See Flax (1983); Rose (1983); Hartsock (1983b); Smith (1974; 1977; 1979; 1981); Harding (1983b); Fee (1981). Haraway (1985) proposes a somewhat different epistemology for feminism.

possible for people to see the world in an enlarged perspective because they remove the covers and blinders that obscure knowledge and observation."[13] The women's movement produces not only the opportunity for such an enlarged perspective but more women scientists, and they are more likely than men to notice androcentric bias.

This solution to the epistemological paradox is appealing for a number of reasons, not the least because it appears to leave unchallenged the existing methodological norms of science. It is easier to gain acceptance of feminist claims through this kind of argument, for it identifies only bad science as the problem, not science-as-usual.

Its considerable strategic advantage, however, often leads its defenders to overlook the fact that the feminist empiricist solution in fact deeply subverts empiricism. The social identity of the inquirer is supposed to be irrelevant to the "goodness" of the results of research. Scientific method is supposed to be capable of eliminating any biases due to the fact that individual researchers are white or black, Chinese or French, men or women. But feminist empiricism argues that women (or feminists, whether men or women) *as a group* are more likely to produce unbiased and objective results than are men (or nonfeminists) as a group.

Moreover, though empiricism holds that scientific method is sufficient to account for historical increases in the objectivity of the picture of the world that science presents, one can argue that history shows otherwise. It is movements for social liberation that have most increased the objectivity of science, not the norms of science as they have in fact been practiced, or as philosophers have rationally reconstructed them. Think, for instance, of the effects of the bourgeois revolution of the fifteenth to seventeenth centuries, which produced modern science itself; or of the effects of the proletarian revolution of the nineteenth and early twentieth centuries. Think of the effects on scientific objectivity of the twentieth-century deconstruction of colonialism.

We shall also see that a key origin of androcentric bias can be found in the selection of problems for inquiry, and in the definition of what is problematic about these phenomena. But empiricism insists that its methodological norms are meant to apply only to the "context of justification"—to the testing of hypotheses and interpretation of evidence—not to the "context of discovery" where problems are identified and defined. Thus a powerful source of social bias appears completely

---

[13]Millman and Kanter (1975, vii).

to escape the control of science's methodological norms. Finally, it appears that following the norms of inquiry is exactly what often results in androcentric results.

Thus, feminist attempts to reform what is perceived as bad science bring to our attention deep logical incoherences and what, paradoxically, we can call empirical inadequacies in empiricist epistemologies.

The feminist standpoint originates in Hegel's thinking about the relationship between the master and the slave and in the elaboration of this analysis in the writings of Marx, Engels, and the Hungarian Marxist theorist, G. Lukacs. Briefly, this proposal argues that men's dominating position in social life results in partial and perverse understandings, whereas women's subjugated position provides the possibility of more complete and less perverse understandings. Feminism and the women's movement provide the theory and motivation for inquiry and political struggle that can transform the perspective of women into a "standpoint"—a morally and scientifically preferable grounding for our interpretations and explanations of nature and social life. The feminist critiques of social and natural science, whether expressed by women or by men, are grounded in the universal features of women's experience as understood from the perspective of feminism.[14]

While this attempted solution to the epistemological paradox avoids the problems that beset feminist empiricism, it generates its own tensions. First of all, those wedded to empiricism will be loath to commit themselves to the belief that the social identity of the observer can be an important variable in the potential objectivity of research results. Strategically, this is a less convincing explanation for the greater adequacy of feminist claims for all but the already convinced; it is particularly unlikely to appear plausible to natural scientists or natural science enthusiasts.

Considered on its own terms, the feminist standpoint response raises two further questions. Can there be *a* feminist standpoint if women's (or feminists') social experience is divided by class, race, and culture? Must there be Black and white, working-class and professional-class, American and Nigerian feminist standpoints? This kind of consideration leads to the postmodernist skepticism: "Perhaps 'reality' can have 'a' structure only from the falsely universalizing perspective of the master. That is, only to the extent that one person or group can

---

[14]Flax (1983), Rose (1983), Hartsock (1983b), and Smith (1974; 1977; 1979; 1981) all develop this standpoint approach.

dominate the whole, can 'reality' appear to be governed by one set of rules or be constituted by one privileged set of social relations."[15] Is the feminist standpoint project still too firmly grounded in the historically disastrous alliance between knowledge and power characteristic of the modern epoch? Is it too firmly rooted in a problematic politics of essentialized identities?

Before turning briefly to the feminist postmodernism from which this last criticism emerges, we should note that both of the preceding epistemological approaches appear to assert that objectivity never has been and could not be increased by value-neutrality. Instead, it is commitments to antiauthoritarian, antielitist, participatory, and emancipatory values and projects that increase the objectivity of science. Furthermore, the reader will need to avoid the temptation to leap to relativist understandings of feminist claims. In the first place, feminist inquirers are never saying that sexist and antisexist claims are equally plausible—that it is equally plausible to regard women's situation as primarily biological *and* as primarily social, or to regard "the human" both as identical *and* nonidentical with "the masculine." The *evidence* for feminist vs. nonfeminist claims may be inconclusive in some cases, and many feminist claims that today appear evidentially secure will no doubt be abandoned as additional evidence is gathered and better hypotheses and concepts are constructed. Indeed, there should be no doubt that these normal conditions of research hold for many feminist claims. But agnosticism and recognition of the hypothetical character of all scientific claims are quite different epistemological stances from relativism. Moreover, whether or not feminists take a relativist stance, it is hard to imagine a coherent defense of cognitive relativism when one thinks of the conflicting claims.

Feminist postmodernism challenges the assumptions upon which feminist empiricism and the feminist standpoint are based, although strains of postmodernist skepticism appear in the thought of these theorists, too. Along with such mainstream thinkers as Nietzsche, Derrida, Foucault, Lacan, Rorty, Cavell, Feyerabend, Gadamer, Wittgenstein, and Unger, and such intellectual movements as semiotics, deconstruction, psychoanalysis, structuralism, archeology/genealogy, and nihilism, feminists "share a profound skepticism regarding universal (or univ-

---

[15]Flax (1986, 17). Strains of postmodernism appear in all of the standpoint thinking. Of this group, Flax has most overtly articulated also the postmodernist epistemological issues.

ersalizing) claims about the existence, nature and powers of reason, progress, science, language and the 'subject/self.' "[16]

This approach requires embracing as a fruitful grounding for inquiry the fractured identities modern life creates: Black-feminist, socialist-feminist, women-of-color, and so on. It requires seeking a solidarity in our oppositions to the dangerous fiction of the naturalized, essentialized, uniquely "human" (read "manly") and to the distortion and exploitation perpetrated on behalf of this fiction. It may require rejecting fantasized returns to the primal wholeness of infancy, preclass societies, or pregender "unitary" consciousnesses of the species—all of which have motivated standpoint epistemologies. From this perspective, feminist claims are more plausible and less distorting only insofar as they are grounded in a solidarity between these modern fractured identities and between the politics they create.

Feminist postmodernism creates its own tensions. In what ways does it, like the empiricist and standpoint epistemologies, reveal incoherences in its parental mainstream discourse? Can we afford to give up the necessity of trying to provide "one, true, feminist story of reality" in the face of the deep alliances between science and sexist, racist, classist, and imperialist social projects?

Clearly, there are contradictory tendencies among the feminist epistemological discourses, and each has its own set of problems. The contradictions and problems do not originate in the feminist discourses, however, but reflect the disarray in mainstream epistemologies and philosophies of science since the mid–1960s. They also reflect shifting configurations of gender, race, and class—both the analytic categories and the lived realities. New social groups—such as feminists who are seeking to bridge a gap between their own social experience and the available theoretical frameworks—are more likely to hone in on "subjugated knowledge" about the world than are groups whose experience more comfortably fits familiar conceptual schemes. Most likely, the feminist entrance into these disputes should be seen as making significant contributions to clarifying the nature and implications of paradoxical tendencies in contemporary intellectual and social life.

The feminist criticisms of science have produced an array of conceptual questions that threaten both our cultural identity as a demo-

---

[16]Flax (1986, 3). This is Flax's list of the mainstream postmodernist thinkers and movements. See Haraway (1985), Marks and de Courtivron (1980), and *Signs* (1981) for discussion of the feminist postmodernist issues.

cratic and socially progressive society and our core personal identities as gender-distinct individuals. I do not mean to overwhelm these illuminating lines of inquiry with criticisms so early in my study—to suggest that they are not really feminist or that they have not advanced our understanding. On the contrary, each has greatly enhanced our ability to grasp the extent of androcentrism in science. Collectively, they have made it possible for us to formulate new questions about science.

It is a virtue of these critiques that they quickly bring to our attention the socially damaging incoherences in all the nonfeminist discourses. Considered in the sequence described in this chapter, they move us from the Woman Question in science to the more radical Science Question in feminism. Where the first three kinds of criticism primarily ask how women can be more equitably treated within and by science, the last two ask how a science apparently so deeply involved in distinctively masculine projects can possibly be used for emancipatory ends. Where the Woman Question critiques still conceptualize the scientific enterprise we have as redeemable, as reformable, the Science Question critiques appear skeptical that we can locate anything morally and politically worth redeeming or reforming in the scientific world view, its underlying epistemology, or the practices these legitimate.

2 GENDER AND SCIENCE: TWO PROBLEMATIC CONCEPTS

Feminist critics face immense obstacles in trying to construct a theory of gender as an analytic category that is relevant to the natural sciences. These obstacles have their origins not only in familiar but inadequate notions of gender but also in certain dogmatic views about science toward which even feminists are often insufficiently critical.

OBSTACLES TO THEORIZING GENDER

In such other disciplines as history, anthropology, and literature, the need to theorize gender appeared only after the limitations of three other projects were recognized. The "woman worthies" project was concerned with restoring and adding to the canons the voices of significant women in history, novelists, poets, artists, and so forth. Their achievements were reevaluated from a nonsexist perspective. The "women's contributions" project focused on women's participation in activities that had already appeared as focuses of analysis in these disciplines—in abolition and temperance struggles, in "gathering" activities within so-called hunter cultures, in the work of significant literary circles, for instance—but were still misperceived and under-developed subject matters. Here, the goal of a less distorted picture of social life logically called for new accounts of these already acknowledged disciplinary subject matters. Finally, "victimology" studies documented the previously ignored or misogynistically described histories and present practices of rape, wife abuse, prostitution, incest, workplace discrimination, economic exploitation, and the like.

30

It was only in doing such work effectively that feminist scholars came to recognize the inadequacy of these approaches. The situation of women who managed to become significant figures in history or recognized artists and poets was by definition privileged in comparison with women's situation more generally. The lives of these women offer us little more understanding of the daily lives of the vast majority of women than lives of great men reveal the lot of the "common man." Furthermore, women's contributions to traditional history and culture have still been contributions to what men, from the perspective of their lives, think of as history and culture. Such analyses tend to hide from us what women's activities in these men's worlds meant to women, as well as how women's daily activities have shaped men's very definitions of their worlds.[1] Finally, the victimology studies often hide the ways in which women have struggled against misogyny and exploitation. Women have been active agents in their own destinies—even if not within conditions of their own making—and we need to understand the forms and focuses of their struggles. These three kinds of studies have all provided valuable insights into matters that traditional inquiry bypasses. But their limitations led feminists to see the need to formulate gender as a theoretical category, as the analytic tool through which the division of social experience along gender lines tends to give men and women different conceptions of themselves, their activities and beliefs, and the world around them.

In the natural sciences, these projects have been only marginally useful. Women have been more systematically excluded from doing serious science than from performing any other social activity except, perhaps, frontline warfare. The inevitable examples of Marie Curie and now Barbara McClintock notwithstanding, few women have been able to achieve eminence in their own day as scientists. A variety of historical, sociological, and psychological studies explain why this is so, but the fact remains that there are few woman worthies to restore to science's halls of fame. Studies of women's contributions to science have been somewhat more fruitful though still limited by the same constraints.[2] The victimology focus, which appears in all five of the feminist science critique projects, has proved valuable chiefly in exploding the myth that the science we have had actually is the "science

[1]See, e.g., Smith (1974; 1977; 1979; 1981); Kelly-Gadol (1976); Gilligan (1972).
[2]See, e.g., Rossiter (1982b); Walsh (1977).

31

for the people" (Galileo's phrase) imagined at the emergence of modern science.

The fact that these approaches, useful in the social sciences and humanities, have been able to find only limited targets in the natural sciences has obscured to the science critics the need for more adequate theorization of gender as an analytic category—with one important exception: in the critiques of biology, there have been great advances in providing more developed and accurate views of women's natures and activities (see Chapter 4). Here the need to theorize gender as an analytic category can be seen in identifications of a gap between the way men and women think about reproduction and reproductive technologies, in questions about whether sex difference itself is not an issue of interest more to men than to women, in suggestions that scientific method's focus on differences might be implicated in the androcentrism of such problematics, and in proposals that the concern in biology, anthropology, and psychology with interactive relationships between organisms, and between organisms and environments, may reflect a specifically feminine way of conceptualizing very abstract relationships.[3]

But biology is only one of the focuses of the feminist critiques of science. In general, the areas in which there is a need for gender as an analytic category and the directions such theorizing should take still remain obscure to many feminist critics of natural science, and totally incomprehensible to most nonfeminist scientists as well as historians, sociologists, and philosophers of science. At least some of these critics do have the resources of their social science disciplines and of literary criticism with which to try to understand natural science in terms of gender categories. The methods of psychoanalysis, history, sociology, anthropology, political theory, and literary criticism have produced valuable insights; however, scientific training (and I include training in the philosophy of science) is hostile to these methods of seeking knowledge about social life, and gender theory is a theory about social

---

[3]However, these suggestions raise as many questions as they answer. For instance, does not this approach tend to universalize the feminine, and thereby reinforce problematic modernist tendencies in feminism toward a politics (and epistemology) based on identities rather than solidarities? And are not interactive models the obvious alternative to the hierarchical models of Darwinian dogma? That is, do not reasons internal to the logic of theory development suggest the fruitfulness of pursuing interactive models at this moment in the history of the biological sciences? Furthermore, does not the desire to replace hierarchical with interactive models reflect widely recognized political realities at this time in world history, rather than only feminine characteristics? We shall pursue these questions later.

life. Characteristically, neither scientists nor philosophers of science are socialized to value psychoanalysis, literary criticism, or the critical interpretive approaches to be found in history and anthropology as modes of knowledge-seeking. No wonder we have found it difficult to theorize the effects on the natural sciences of gender symbolism, gender structure, and individual gender.

In the social sciences, those areas of research most hospitable to the introduction of gender as a theoretical category are the ones with a strong *critical* interpretive tradition. (I say "critical" to distinguish this *theory* of human action and belief from the kinds of unselfconscious interpretations, rationalizations, we all routinely provide to ourselves and others in explaining our beliefs and actions.) These traditions hypothesize that "the natives" may sometimes engage in irrational actions and hold irrational beliefs that defeat the actors' conscious goals and/or unconscious interests. The causes are to be found in the contradictory social conditions, the no-win situations, within which humans must choose actions and hold beliefs. Marx and Freud provide just two examples of theorists who attempted to identify the social conditions that lead groups of individuals to patterns of irrational action and belief. The effects of their methodological proposals can be seen in the critical interpretive traditions in many areas of social science research—whether or not these traditions call themselves Marxist or Freudian or are concerned with the particular kinds of social phenomena of interest to Marx and Freud. In these inquiry traditions it is legitimate—indeed, often obligatory—to reflect on the social origins of conceptual systems and patterns of behavior, and to include in this subject matter the conceptual systems and behaviors shaping the inquirer's own assumptions and activities. Here there is not only conceptual space but also, we might say, moral permission to reflect on gendered aspects of conceptual systems and on the gender circumstances in which beliefs are adopted. In contrast, research programs where remnants of empiricist, positivist philosophies of social science hold sway have been systematically inhospitable to gender as a theoretical category.[4] At best they have been willing to add gender as a

---

[4] See Stacey and Thorne (1986), who make a number of these points about sociology. Pauline Bart has also pointed out (in conversation) that in speculating about the comparative resistances that different disciplinary fields offer to feminist insights, we should not underestimate the comparative levels of personal and political threat to the leaders in these fields—primarily men—that are presented, for instance, by sociological analyses of contemporary and nearby cultures in comparison to historical or anthropological analyses of cultures temporally or spatially distant from us. This line of reasoning

variable to be analyzed in their subject matter—as a property of in-dividuals and their behaviors rather than also of social structures and conceptual systems.

The physical sciences are the origin of this positivist, excessively empiricist philosophy. Their nonsocial subject matter and the para-digmatic status of their methods appear to preclude critical reflection on social influences on their conceptual systems; indeed, prevalent dogma holds that it is the virtue of modern science to make such reflection unnecessary. We are told that modern physics and chemistry eliminate the anthropomorphizing characteristic of medieval science and of the theorizing we can observe in "primitive" cultures and chil-dren—not to mention in the social sciences and humanities. The social progressiveness, the "positivism," of modern science is to be found entirely in its method. There is thought to be no need to train phy-sicists, chemists, or biologists as critical theorists; consequently, little in their training or in the ethos of scientific endeavor encourages the development or appreciation of the critical interpretive theory and skills that have proved so fruitful in the social sciences.

However, the history, sociology, and philosophy of science are not themselves natural sciences. Their subject matters are social beliefs and practices. In the philosophy of science, the focus is on ideal beliefs and practices; in the history and sociology of science it is on actual beliefs and practices. Whether ideal or real, social beliefs and practices are the concerns of these disciplines. Here one would have thought that critical interpretive theory and skills would be central to under-standing how scientists do and should explain the regularities of nature and their underlying causal tendencies. The sociology of knowledge does take this approach, though it has been limited by its preoccupation with what we can call the "sociology of error" and the "sociology of knowers" to the exclusion of a sociology of knowledge.[5] And this tradition, too, has been stalwartly androcentric. But androcentric or not, its influence on thinking about natural science has yet to be felt within the philosophy of science or the natural sciences themselves, and is only beginning to make inroads into the traditional sociology and history of science. The philosophy, sociology, and history of the natural sciences have been dominated by empiricist philosophies hostile

would support my argument that feminist critiques of the natural sciences meet even greater hostility than critiques in other areas; scientific rationality is directly implicated in the maintenance of masculinity in our kind of culture.

[5] See Bloor (1977) for criticism of the sociologies of error and knowers.

to theories of belief formation within which gender could be understood as a part of science's conceptual schemes, as a way of organizing the social labor of science, or as an aspect of the individual identity of scientists.

For these reasons the feminist science critics face even greater disciplinary obstacles than do feminists who seek to introduce gender as a theoretical category into the social sciences, literature, and the arts. These obstacles seem to originate in the unusual notion that science enthusiasts have of the proper way to understand the history and practices of science: this kind of social activity alone, we are told, must be understood only in terms of its enthusiasts' understanding of their own activities—in terms of the unselfconscious, uncritical interpretations "the natives" provide of their beliefs and activities. That is, scientists report their activities, and philosophers and historians of science interpret these reports so that we can "rationally" account for the growth of scientific knowledge in the very same moral, political, and epistemological terms scientists use to explain their activities to funding sources or science critics.

Social theorists will recognize this approach as a hermeneutic, intentionalist one that systematically avoids critical examination of the identifiable causal, historical influences on the growth of science which are to be found outside the intellectual, moral, and political consciousnesses of science practitioners and enthusiasts.[6] Kuhn's alternative account of the history of science has generated a veritable new industry for the social studies of science, studies that have begun to show the mystification perpetrated by such "rational reconstructions."[7] But traditional science and philosophical and popular enthusiasm for the traditional vision of science remain pugnaciously hostile to such critical causal accounts. From this perspective, my approach to science in this book may be understood as a more thorough naturalism than science enthusiasts themselves are apparently willing to defend: I seek to identify the causal tendencies in social life that leave traces of gender projects on all aspects of the scientific enterprise.

Is it ironic that natural science, presented as the paradigm of critical, rational thinking, tries to suffocate just the kind of critical, rational thought about its own nature and projects that it insists we must exercise about other social enterprises? Perhaps not, if we think of

[6]See Fay and Moon (1977) for discussion of the virtues and problems of intentionalist approaches to social inquiry.
[7]Kuhn (1970).

35

science's story about itself as a kind of origins myth. Science's self-image presents a myth about who "our kind" of people are and about what destiny nature and scientific rationality hold in store for us. As anthropologists tell us, origins myths frequently violate the very categories they generate: in other cultures they may report that those cultures came into existence through incest, cannibalism, bestiality, sexual unions between gods and mortals—activities subsequently forbidden in those cultures. The origins myth for our scientific culture tells us that we came into existence in part through the kind of critical thought about the social relations between medieval inquiry and society that is subsequently forbidden in our scientific culture. This is a magical—perhaps even a religious or mystical—conception of ideal knowledge-seeking. It excludes itself from the categories and activities it prescribes for everything else. It recommends that we understand everything but science through causal analyses and critical scrutiny of inherited beliefs.

THE DOGMAS OF EMPIRICISM

Empiricist conceptions of scientific method and the scientific enterprise create obstacles both *for* and *in* feminist thinking about science. I suggest that we should regard these mystifying beliefs as reflections of and additions to the "dogmas of empiricism" familiar to philosophers.

In the 1950s, the philosopher of science Willard Van Orman Quine identified two dogmas of empiricism that he thought should be abandoned. "Modern empiricism has been conditioned in large part by two dogmas. One is a belief in some fundamental cleavage between truths which are *analytic*, or grounded in meanings independently of matters of fact, and truths which are *synthetic*, or grounded in fact. The other dogma is *reductionism*: the belief that each meaningful statement is equivalent to some logical construct upon terms which refer to immediate experience."[8] Quine argued that both dogmas were illfounded, and that if they were abandoned, we would be inclined to see as less clear the purportedly firm distinction between natural science and speculative metaphysics. We would also recognize pragmatic standards as the best we can have for judging the adequacy of scientific claims.

Since then, historians and sociologists of science as well as philosophers have supported Quine's rejection of these two dogmas of em-

[8]Quine (1953, 20).

piricism. Studies of the social construction of what we count as real—both inside and outside the history of science—make it highly implausible to believe that there can be any kind of value-free descriptions of immediate experience to which our knowledge claims can be "reduced" or thought equivalent. Furthermore, there is now widespread acceptance of Quine's first claim that when epistemological push comes to shove, we can never tell for sure when we are responding to the compulsions of our language rather than to those of our experience. Facts cannot be separated from their meanings. Thus the test of the logical adequacy of a statement or argument is ultimately not different in kind from tests of its empirical adequacy. In both cases, (social) experience expressed through (culturally shaped) language is all we have to fall back on. (Quine was not concerned with what creates social variation in experience or language.) Quine recommended substituting pragmatic and behaviorist questions for the traditional philosophical ones, replacing what he thought were undesirable philosophical preoccupations with what he thought were desirable scientific ones. We can appreciate the pragmatic tendencies in his thinking without having to agree to his behaviorism—to his program for replacing philosophy with what appears to many theorists as a still far too reductionist and obsessively empiricist social science.

The philosophical preoccupations that concerned Quine were developed in their contemporary forms to explain the emergence of modern science;[9] philosophers and scientists explicitly honored those dogmas. However, both the resistance of the natural sciences to a feminist critique and the many theoretical and political contradictions within the feminist critiques make clear that by no means have the dogmas Quine identified been abandoned—nor are there only two—in either scholarly or popular thinking about science.

Here I want to discuss a series of reflections of and additions to the assumptions Quine criticized which stand as conceptual obstacles to our ability to analyze science, too, as a fully social activity. I think these excessively empiricist beliefs still haunt most of the feminist critics of science and prevent us from adequately theorizing gender in feminist discussions of science. Furthermore, it is belief in these dogmas that leads scientists and traditional philosophers and historians to be hostile to the very idea of a feminist science critique.

[9]See Rorty (1979).

Sacred Science.

I have already hinted at one of these dogmas: the belief that science is a fundamentally unique kind of social activity. Like other kinds of origins stories, the ideology of science claims that science properly violates the categories it generates. We are told that human understanding is decreased rather than increased by attempting to account for the nature and structure of scientific activity in the ways science recommends accounting for all other social activity. This belief makes science sacred. Perhaps it even removes scientists from the realm of the completely human—at least in their own view and the view of science enthusiasts. It sets limits on human rationality for what are best thought of as religious or mystical reasons.

We can illustrate that the problem lies in inadequate conceptions of scientific rationality rather than in specifically feminist claims by considering the following hypotheses—which do not even refer to gender.

A. The predictable contribution that physics could make to social welfare today is relatively negligible, since moral and political injustices, rather than ignorance of the laws of nature, are the greatest obstacles to social welfare.

B. "More science" in a socially stratified society tends to intensify social stratification.

C. While individual scientists may well be motivated by the loftiest of personal goals and social ideals, their current activity in fact functions primarily to increase profit for and maintain social control by the few over the many.

These claims may be true or false; I think they are closer to truth than to falsity. Determining their truth or falsity—their correspondence with the way the world is—should be considered a matter for empirical investigation. Yet these statements appear blasphemous to the vast majority of both scientists and nonscientists—not bold hypotheses that should be scientifically investigated to determine whether or not they can be refuted but psychologically, morally, and politically threatening challenges to the Western faith in progress through increased empirical knowledge. They also appear as challenges to the intelligence and morals of the very bright and well-intentioned women and men who enter and remain in science. The usual responses to such suggestions are raised eyebrows, knowing smiles (not directed toward the speaker), or overtly hostile glares—responses that are hardly paradigms of rational argument. Alternatively, listeners may indicate that they think they are hearing simply expressions of personal hurt: "You must hate scientists," they reply—as if only disastrous personal experience or a warped mind could make such hypotheses worth pursuing. These kinds of statements raise the possibility not just of an interesting empirical discovery that we have been in error about the progressiveness of science today but of a painful, world-shattering confrontation with moral and political values inconsistent with those that most people think give Western social life its desirable momentum and direction. Obviously, more is at issue here than checking hypotheses against facts—just as more was at issue in the social acceptance of the Copernican world view than the relationship between Copernicus's hypotheses and the evidence to be gained by looking through Galileo's telescope.

The project that science's sacredness makes taboo is the examination of science in just the ways any other institution or set of social practices can be examined. If one substituted "novels," "drama," "marriage," or "publicly funded education" for "science" in these claims, many people might be outraged (or consider the claims merely silly), but the hypotheses would not then generate the same deep feeling of threat to our moral, political, and psychological intuitions. Why is it taboo to suggest that natural science, too, is a social activity, a historically varying set of social practices? that a *thoroughgoing* and *scientific* appreciation of science requires descriptions and explanations of the regularities and underlying causal tendencies of science's own social practices and beliefs? that scientists and science enthusiasts may have the least adequate understanding of the real causes and meanings of their own activities? To what other "community of natives" would we give the final word about the causes, consequences, and social meanings of their own beliefs and institutions? If we are not willing to try to see the favored intellectual structures and practices of science as cultural artifacts rather than as sacred commandments handed down to humanity at the birth of modern science, then it will be hard to understand how gender symbolism, the gendered social structure of science, and the masculine identities and behaviors of individual scientists have left their marks on the problematics, concepts, theories, methods, interpretations, ethics, meanings, and goals of science.

Let us pursue for a moment the way this belief in the sacredness of science is defended. Science and society are analytically separate, we are told. Thus social values are distinct from (and detrimental to the determination of) facts; the meanings scientific statements carry in a

39

culture are distinct from (and irrelevant to) what scientific statements actually say; consideration of the social uses and abuses of science are distinct from (and irrelevant to) assessments of the progressiveness of science; the social origins of scientific problematics, concepts, theories are distinct from (and irrelevant to) the "goodness" of these problematics, concepts, and theories. These beliefs are defended in one form or another every time a social criticism of science appears. Furthermore, these beliefs permit continual discussions in which the languages, meanings, and structures of science are assumed to be uniquely asocial, as a quick perusal of any of the standard philosophy of science journals or texts will reveal. These beliefs structure the internalist vs. externalist dispute in the history of science; they ensure that most science enthusiasts will mean by "history of science" only the history of consciously held scientific beliefs.

Defenders of the analytic separateness of science from society will say that maybe science is not immune from *all* kinds of social influences; anyone can see that idiosyncrasies of individual investigators have influenced the history of science—otherwise, why would we give Nobel prizes to some individuals and not to others? And yes, the funding priorities of the economy and state do influence the selection of problematics. And it's also true that shoddy research sometimes survives longer than it should because of social enthusiasm for the ill-begotten interpretations of its results: think of Lysenkoism and "Nazi science," they say. And of course enthusiasm for modern science is fundamentally motivated by democratic social values: science is constituted by certain social values, but at its best it neither defends nor recommends any particular social values.

What the defenders of the fundamental value-neutrality, the purity, of science really mean, they say, is that science's logic and methodology, and the empirical core of scientific facts these produce, are totally immune from social influences; that logic and scientific method will in the long run winnow out the factual from the social in the results of scientific research. But we shall try to locate the pure, value-free core of science responsible for the purportedly inherent progressiveness in scientific method, in model claims in physics, in the mathematical language of science, and in logical reasoning. If, as I shall argue, pure science cannot be found in these places, then where should we try to find it?

We do know where to find the historical origins of the mystical belief that science's inherent progressiveness resides in the separation of its logic and its facts from its social origins, social uses, and social meanings; Chapter 9 examines the political reasons for its adoption. Prior to Newton, such a positivist view of science did not exist (though the term "positivism" appeared much later, the idea can already be detected in late seventeenth-century thinking). The separation does not in fact exist today, but its fetishization lingers on.

Science as a Unique Method or a Set of Sentences.
Does the feminist case that science is gendered have to rest on showing scientific method to be sexist? Does a degendered science have to produce a new method of knowledge-seeking? Or does the feminist case have to rest on showing that the best confirmed claims the sciences have made are sexist? Does it have to show that Newton's or Einstein's laws are sexist in order to provide a plausible argument for the gendered nature of science?

The common view (or dogma) is that science's uniqueness is to be found in its method for acquiring reliable descriptions and explanations of nature's regularities and their underlying causes. Authors of science texts write about the importance of value-free observation as the test of beliefs, and especially about collecting observations through the "experimental method." We are told that it is the refined observation characteristic of experimental method that permitted Galileo's and Newton's views to win out over Ptolemy's and Aristotle's.

But exactly what is unique about this method remains obscure. For one thing, the different sciences use different methods; not a great deal is common to the methods of astronomy, particle physics, and molecular biology. For another thing, in parts of what are regarded as highly rigorous and value-free sciences—contemporary astronomy and geology, for example—controlled experiment plays an extremely small role. And controlled experiment is not a modern invention—after all, Aristotle was an experimentalist. Moreover, just try to identify the formal methodological features of knowledge-seeking that will exclude from the ranks of scientists farmers in premodern agricultural societies yet will include junior but highly trained members of biochemical research teams. When push comes to shove in the philosophy of science, we are told that induction and deduction are supposed to compete for honors as the core of scientific method.[10] But presumably, human infants as well as apes and dogs regularly use induction and deduction.

[10]Popper (1959; 1972); cf. Harding (1976).

These kinds of considerations lead to the suspicion that science is both more and less than any possible definition of scientific method.

Faced with these kinds of arguments, one leading philosopher of science says that what distinguishes scientific from nonscientific explanation is science's *attitude* toward its claims.[11] That is, what makes a belief or activity scientific is the psychological stance one takes toward it. In all other kinds of human knowledge-seeking, we can identify assumptions that are regarded as sacred, as immune from refutation by experience; the explanations offered by non-Western, "primitive" cultures, theology, psychoanalytic theory, Marxist political economy and astrology are the favorite examples of such pseudoexplanations. We are told that only science holds all of its beliefs open to refutation by experience.

However, in particular areas of scientific inquiry the immunity to criticism of grounding assumptions is easily demonstrated. Why should the situation be different for the scientific world view as a whole? How about (one is tempted to ask) the belief that there are no uncaused physical events? Or that we can meaningfully distinguish between the world's physical and nonphysical events or processes?

In light of these kinds of considerations, it is hard to see why a distinctively feminist science would have to produce a new method, at least if we mean by scientific method no more than (1) putting beliefs to the test of experimental observation, (2) relying on induction and deduction, or (3) being willing to hold all of our assumptions open to criticism. The first and second of these activities are not at all unique to modern science, and the third is not characteristic of what everyone counts as the most methodologically rigorous inquiry. What we have in this dogma is the reduction of the purportedly inherent progressiveness of science to a mythologized and obscure notion of its method (this should be—but is not always—what feminists criticize when they challenge positivism), but the distinguishing features of this scientific method cannot even be specified in a plausible way.

A second obscuring conception can be found in the history of the philosophical and scientific preoccupation with science as a particular paradigmatic set of sentences. The mathematical expressions of Newton's laws of mechanics or Einstein's theory of relativity are two of the most frequently cited examples. Unless critics can show that these mathematical statements are value-laden, it is claimed, no case at all

[11]Popper (1959; 1972).

can be made for the hypothesis that the science we have is fundamentally suffused with social values—let alone with gender values. But why should we continue to regard physics as the paradigm of scientific knowledge-seeking? And is it true that mathematical statements bear no social fingerprints—that there is such a thing as pure mathematics?

Paradigmatic Physics.
Physicists, chemists, philosophers of science, and most of the rest of us believe that physics is the paradigm of science, and that science without physics as its paradigm is unimaginable. Minds reel at the suggestion that perhaps, in the science of the future, physics will be relegated to the backwaters of knowledge-seeking and thought to be concerned only with esoteric problems that have little impact on how we live. Perhaps even today its problematics, methods, and favored languages already provide distinctly atypical examples of scientific inquiry that should not be models for other areas. We can entertain this thought even while we appreciate the historical reasons why physics has been the paradigm of scientific inquiry: Newton's physics permitted a far more useful understanding of many kinds of phenomena than did the Aristotelian physics it replaced, and its explanatory success created great optimism that Newton's "method" could produce similar success in every area of human inquiry. Indeed, mechanism, the metaphysics of Newton's laws, still guides useful research in many areas of the physical sciences, though its limitations are becoming increasingly apparent. However, as Kuhn pointed out, paradigmatic theories in particular areas of inquiry eventually wear out as fruitful guides to research. Shouldn't this also be true for science as a whole?

If it is reasonable to believe that physics should always be the paradigm of science, feminism will not succeed in "proving" that science is as gendered as any other human activity unless it can show that the specific problematics, concepts, theories, language, and methods of modern physics are gender-laden—especially, one hears from philosophers, mathematicians, and physicists, that the mathematical expressions of Newton's laws of mechanics and Einstein's relativity theory are gender-laden. Here, surely, we can distinguish the value-neutral logical structure and empirical content of scientific belief from its social origins, meanings, and applications. From this perspective, the feminist science critiques appear to have as their targets only the "less rigorous" or "less mature" biological and social sciences. Resistance to the plausibility of the feminist critique is made to rest on the value-neutrality of mathematical expressions of the laws of physics. Thus feminist criticisms can appear to support the claim that specific examples of sexist and androcentric science are only cases of "bad science"; that greater attention to the methodological constraints modeled by physics for all inquiry would result in a science free of sexism and androcentrism.

The fact is, however, that all the reasons social scientists have given for thinking that social inquiry requires fundamentally different metaphysical assumptions and methods from those of inquiry in physics can be understood as reasons for thinking that the status of physics as the model of science should deteriorate.[12] I will argue that a critical and self-reflective social science should be the model for all science, and that if there are any special requirements for adequate explanations in physics, they are just that—special. (We will see that much of biology should already be conceptualized as social science. Thought of as the bridge between—or, from a postmodernist perspective, the crucible in which are forged—the natural and the social, nature and culture, biology must frequently make kinds of metaphysical and methodological assumptions that are foreign to physics and chemistry.) Let us see how the arguments about the different conditions for adequate social inquiry can be transformed into arguments for regarding the conditions of scientific explanation in physics as nonparadigmatic.

In the first place, the subject matter of physics is so much less complex than the subject matters of biology and the social sciences that the difference amounts to a qualitative rather than just a quantitative one. Physics looks at either simple systems or simple aspects of complex systems. The standard model of the solar system is an example of the former; the aspects of physiological or ecological systems that physics can explain are examples of the latter. A major reason for the simplicity of these systems and the ability of their models to make reliable predictions is that they are conceptualized as self-contained and deterministic. Yet human activity can have consequences for the functioning of the solar system—we could, presumably, blow up this planet. But the regularities and causal tendencies of such kinds of "interference" are not supposed to be the professional concern of physicists. Whereas the social sciences must consider physical constraints on the phenomena they examine, the objects, events, and processes of

---

[12]See Fay and Moon (1977) for a review of how mainstream philosophers think about the differences between the physical and social sciences.

concern to physical scientists are limited to those that can be isolated from social constraints.

Second, the concepts and hypotheses of physics require acts of social interpretation no less than do those in the social sciences. The social meanings that explanations in physics have for physicists and for the "man and woman in the street" are necessary components of these explanations, not scientifically irrelevant historical accidents. Perhaps it is appealing to imagine that the mathematical formulations of Newton's laws *are* the explanations of the movements of matter because it takes only a little effort for us modern folk to get a sense of what these formulas mean in ordinary language. But should we think of a formula so long that only a computer could read it in one hour as an *explanation* of a type of phenomenon? The answer to this question is "no." An explanation is a kind of social achievement. A purported explanation that cannot be grasped by a human mind cannot qualify as an explanation. If no human can understand, can hold in the mind, the purported explanation, then explanation has not been achieved. In other words, Newton's explanations include not just the mathematical expressions of his laws but also the interpretations of those formulas that let us know when we have cases in front of us that exemplify the formulas. The formula "$1 + 1 = 2$" is meaningless unless we are told what is to count as a case of 1, of $+$, of $=$, and so on. The history of chemistry can be understood in part as the struggle to determine what should count as the 1's, the $+$'s, and the $=$'s of chemical "addition." And it is not just in physics and chemistry that the appropriate meanings and referents for such apparently obvious terms are debated. As a famous physicist is alleged to have remarked, if we put one lion and one rabbit in a cage, we rarely find two animals there one hour later! Scientific formulas are like legal judgments: the laws become meaningful only through learning (or deciding) how to apply them, and doing so is a process of social interpretation.

We can see another way in which social interpretation is a fundamental component of the laws of physics if we think about the fact that we, unlike fifteenth- to seventeenth-century Europeans, no longer find it bizarre or morally offensive to conceptualize nature as a machine. This analogy has become so deeply embedded in our cultural consciousness that no longer are we aware when we draw on it. But we do not think of concepts or hypotheses "interpreted" through *unfamiliar* social analogies as contributing to explanations. "Nature is like a 'speak bitterness' meeting" might conceptualize nature in a way that could fruitfully guide scientific inquiry in some cultures but not in ours (perhaps Chinese ecologists might find this a useful metaphor). An "explanation" we cannot grasp is not an explanation. A theory's interpretation may overtly appeal to social or political metaphors at one time and not at another, but *some* social act of interpretation is necessary if we are to understand how to use the theory. Interpretation of formal "texts" through socially familiar models and analogies is central to explanations in physics.[13]

In the third place, whereas the evolutionary biologist or economic geographer must take into account purposeful and learned activities by humans and perhaps even members of other species—nonhuman feeding and mating preferences, for example—the physicist need not consider self-reflective and intentionally directed causes of the motions of mere matter. He need not do so because the observable regularities of "matter in motion" do not have these kinds of causes. I mention evolutionary biology and economic geography to indicate how deeply the social extends into what we think of as the natural. After all, explanations of apes' adaptation to (perhaps we should say "creation of") their environments and of patterns of forestation at least since our species came into existence must include considerations of just the kind of purposeful and learned behaviors (dare we say "activities"?) that are the subject matter of social inquiry. Insofar as the world around us continues to become more and more suffused with the presences and residues of social activities, there is less and less "out there" amenable to the kinds of explanations that have been so fruitful in physics. The history of the "progress" of our species is simultaneously the history of the disappearance of pure nature. I need hardly even mention the silliness of assuming that physics can provide the model for anthropological explanations of all we want to know about the regularities and underlying causal tendencies creating different kinds of kinship structures, or for historical explanations of all we want to know about the regularities and underlying causal tendencies in relationships between, say, forms of child rearing and forms of the state. I suggest that the totally reasonable exclusion of intentional and learned behaviors from the subject matter of physics is a good reason to regard inquiry in physics as atypical of scientific knowledge-seeking.

Finally, explaining social phenomena requires the interpretive skills

[13]Later (esp. Chapter 9) I examine the use of androcentric metaphors, models, and analogies in the history of Western science, and the inadequate account of the nature and functions of these figures of thought in the philosophy of science.

necessary to grasp the meanings and purposes an intentional act has for the actor—skills that have no analogue in physics. Indeed, the differences between the ontological assumptions and methods appropriate for physics and social inquiry are even more extensive than such a statement indicates. In social inquiry we also want to explain the origins, forms, and prevalence of apparently irrational but culturewide patterns of human belief and action. Freud, Marx, and many later social theorists have taken just such culturewide irrationality as their subject matter. Why, then, should we take as the model for all knowledge-seeking a science that has no conceptual space for considering irrational behavior and belief? Moreover, possibly explanations even in physics would be more reliable, more fruitful, if physicists were trained to examine critically the social origins and often irrational social implications of their conceptual systems. For instance, would not physics benefit from asking why a scientific world view with physics as its paradigm excludes the history of physics from its recommendation that we seek critical causal explanations of everything in the world around us? Only if we insist that science is analytically separate from social life can we maintain the fiction that explanations of irrational social belief and behavior could not ever, even in principle, increase our understanding of the world physics explains.

I have been suggesting reasons for reevaluating the assumption that physics should be the paradigm of scientific knowledge-seeking. If physics ought not to have this status, then feminists need not "prove" that Newton's laws of mechanics or Einstein's relativity theory are value-laden in order to make the case that the science we have is suffused with the consequences of gender symbolism, gender structure, and gender identity. Instead, we should regard physics as simply the far end of the continuum of value-laden inquiry traditions. Even though there are good historical reasons why physics gained such a central position in the thinking of philosophers and scientists, we need to ask whether its paradigmatic status today should be regarded as anachronistic, and as a reflection of distinctively androcentric, bourgeois, and Western concerns.

Let me emphasize that I do not intend to direct attention away from attempts to show how Newton's and Einstein's laws of nature might participate in gender symbolization. Improbable as such projects may sound, there is no reason to think them in principle incapable of success. Such successes would make immensely more plausible the feminist claims that the natural sciences, too, are deeply gender-biased.

In Chapters 5, 8, and 9, in examining some of the androcentric and bourgeois social values that have in fact been projected onto nature, I will show that modern astronomy and physics anthropomorphize nature no less than did the medieval sciences they replaced. But here I am making a different point. I am arguing that such a project need not be undertaken in order to convince us that modern science is androcentric. Instead, we should understand physics not as the model for all scientific inquiry, but as atypical of inquiry just insofar as its ontological and methodological assumptions can in fact secure value-free results of research.

Pure Mathematics.

The belief that mathematics has no formal social dimensions—that the "external" social history of mathematics has left no traces on its "internal" intellectual structures—provides grounds for regarding science as fundamentally a set of sentences (such as Newton's laws) and physics as the paradigmatic science. For if the nature that modern physics describes and explains "speaks in the language of mathematics" (as Galileo claimed), and if the cognitive content of mathematics has no social characteristics, then the formal statements of physics must also have no social characteristics. We have already argued that explanations in physics cannot be "reduced" to mathematical "sentences" shorn of social interpretation. But the dogmatists' case for a value-neutral core of pure science is even weaker than that argument suggests. Even if one could "reduce" the laws of physics to mathematical expressions, there are not sufficient reasons to think that those mathematical expressions themselves are value-free.

Of course, everyone knows that the field of mathematical inquiry has a social history. Different mathematical problems preoccupied different historical groups of mathematicians. We are told that different concepts, calculation strategies, and methods of proof were "discovered" at identifiable historical moments. But we are also told that this social history of mathematics is entirely external to the cognitive structures, the logical structures, of mathematics. The social history of mathematics is said to leave no traces on its logical structures. These "discoveries" are presented as merely examples of the always cumulative and progressive growth of mathematical knowledge.

It is sometimes claimed that if feminism is to show the value of using gender as a category to analyze science, it must show that mathematical concepts and methods of proof are androcentric, and it must produce an alternative, feminist mathematics; perhaps feminists must even show that modern logic is sexist and that there could be a nonsexist alternative logic. This argument satisfies its makers that they have reduced to an absurdity both the very idea of a radical feminist critique of the scientific world view and the possibility of an alternative science guided by feminist principles.

I will not argue that mathematics is, in fact, *male*-biased; but two considerations make it plausible to regard as mythical the possibility of *pure* mathematics. In the first place, no conceptual system can provide the justificatory grounds for itself. To avoid vicious circularity, justificatory grounds always must be found outside the conceptual system one is trying to justify. The axioms of mathematics are no exception to this rule. Leading mathematical theorists point out that the ultimate test of the adequacy of a mathematical concept or proof always has been pragmatic: Does it "work" to explain the regularities in the world for which it was intended to provide an explanation? The history of the last two centuries of the philosophy of mathematics can be seen as the history of the struggle to arrive at this pragmatic understanding of the nature of mathematical "truths." Our interests here do not permit a review of this history.[14] But on the basis of this now widespread (if not totally convincing to all mathematicians) understanding of the status of mathematical "truths," we should think of "discoveries" in the history of mathematics as responses to the recognition that mathematical concepts and theories, too, are tested against the historical social worlds they are designed to explain.

In the second place, in support of this kind of argument, historians of mathematics have pointed to the reasons why mathematical statements regarded as true at one time in history are occasionally regarded as false at a later time. They show that the plausibility or usefulness of what have sometimes appeared as impossible, contradictory, mathematical concepts has had to be socially negotiated.[15] One kind of social imagery for thinking about mathematical objects comes to replace another. For example, the ancient Greeks—no mean mathematicians—did not regard one, the first in a series of integers, as a number, nor did they consider it either odd or even. We, of course, think of it as

---

[14]See the accounts provided by Kline (1980) and Bloor (1977). Kline argues that Andrzej Mostowski, Hermann Weyl, Haskell B. Curry, John von Neumann, Bertrand Russell, Kurt Gödel, and Quine are among the eminent mathematicians and logicians who have defended a pragmatic view of mathematical truth.

[15]See Bloor (1977) for discussion of these cases.

a number, and as an odd number, because unlike the ancient Greeks, we are not mathematically interested in the distinction between the first, or generator, of a lineage (here of integers) and the lineage generated. Theologies and origins stories frequently invoke such a distinction. In mathematics, we have come to see the distinction between the generator of a lineage and the lineage generated as a distinction originating in certain kinds of social beliefs that modern mathematics need not honor. (However, scientists and philosophers who insist that science itself in principle cannot have some of the characteristics possessed by the world that science explains—illumination by causal explanation, social values in the explanatory artifacts physicists produce, and the like—still retain belief in the importance of this kind of distinction, as I noted. If we no longer can find reasons to honor this religious distinction in mathematics, why should we honor it in the philosophy and social studies of science?)

Let us consider one more example. Common sense tells us that a part cannot be equal to the whole. Thus it is only relatively recently that mathematicians have been able to countenance the idea that the integers could be infinite in number. Earlier mathematicians' problem was as follows: one can match each sequential integer with an even integer (1–2, 2–4, 3–6, 4–8, . . .), resulting in an infinite series in which there are as many even integers as there are integers—at first glance an absurdity. How was this paradox resolved? Mathematicians were willing to let go of the common sense truth that a part cannot be equal to the whole for this special circumstance in order to develop infinitesimal theory. They did so by replacing the social image of numbers as counting units with the social image of numbers as divisions of a line. These are *social* images because they reflect what people in historical cultures intentionally do. Not all cultures have been as preoccupied with measuring—dividing a line—as has ours for the last few centuries. A whole field of mathematical inquiry was made possible by the substitution of a different kind of social image for thinking about what numbers are. As one commentator points out, such a process of socially negotiating cultural images in mathematics is similar to what we do when we exclude patriotic killing in wartime from the moral and legal category of murder.[16]

We could look at these developments in mathematics simply as the onward and upward march of truth in the service of intellectual prog-

---

[16]Bloor (1977, 127). Frances Hanckel's comments improved this discussion.

ress. But to do so hides the social imagery within which numbers and other mathematical notions have been conceptualized, and the very interesting processes of social negotiation through which one cultural image for thinking about mathematical concepts comes to replace another. Counting objects and partitioning a line are common social practices, and these practices can generate contradictory ways of thinking about the objects of mathematical inquiry. It may be hard to imagine what gender practices could have influenced the acceptance of particular concepts in mathematics, but cases such as these show that the possibility cannot be ruled out a priori by the claim that the intellectual, logical content of mathematics is free of all social influence.

"Well, at least mathematics is ultimately grounded by logic; and logic *is* free of social influence," our diehard dogmatist may claim. Mathematicians in this century, however, have found it impossible to justify the axioms of mathematics with any logical principles that are not more dubious, more counterintuitive, than the mathematics they are supposed to justify. So it is doubtful that the duty of providing a firm grounding for the truths of mathematics can be assigned to logic. Moreover, a few feminists have proposed ways in which specific assumptions in logic are androcentric. Merrill Hintikka and Jaakko Hintikka, for example, argue that the metaphysical units of a branch of logic called "formal semantics" correspond to masculine but not feminine ways of individuating objects.[17] Such studies provide invaluable glimpses of social fingerprints on supposedly pure formal thought and suggest fruitful research programs for the future.

But even if these studies did not exist or no more were produced, it is hard to see why the case for theorizing gender as an analytic

---

[17]Hintikka and Hintikka (1983). Another kind of problem in logic was revealed by Janice Moulton in "The Myth of the Neutral 'Man' " in *Feminism and Philosophy*, ed. M. Vetterling-Braggin et al. (Totowa, N.J.: Littlefield Adams, 1977). She pointed out that in a standard English example of a valid syllogistic form—"All men are mortal; Socrates is a man; therefore, Socrates is mortal"—the term "man" in fact is used with two different referents (generic in the first statement; gender-specific in the second), and thus that the standard English interpretation of this syllogism, *used in every logic text for several centuries*, is invalid. The clue to the fact that there are an illicit four, instead of three, terms in this interpreted syllogism is that one can not substitute the name of any and every other "man" (human) for "Socrates" without eliciting a "bizarreness response"; for instance, "Cleopatra is a man" elicits such a response. (The syllogism would, of course, be valid if "men" in the first premise were used in the gender-specific sense; but this does not accurately represent the original Greek, and is not what logicians have intended.) What other androcentric and therefore illicit interpretations of logical forms lurk in logic texts? No wonder many "female men" have had inarticulable resistance to grasping the virtues of logic courses!

51

category in our thinking about science would have to rest on the possibility of producing such analyses of mathematics and logic. My point, again, is not to discourage such studies but to indicate the counterproductiveness (the irrationality!) of this argumentative strategy. This kind of resistance to feminist critiques pays the price of reducing science to mathematical or logical statements, thereby managing to contradict the fundamental assumption that assessments of the adequacy of scientific claims should depend on the detectable relationship of those claims to our observations of the world. It should be sufficient to point out that mathematics is so useful to physics, more limitedly useful in biology or economics, and only rarely useful in anthropology or history because of the relative degrees of simplicity, abstraction, and intentional and irrational behaviors characteristic of the subject matters in these fields of inquiry. Pursuing Quine's turn to pragmatism, we could say that mathematics, like logic, simply "looks at" aspects of the world that are less distorted by formal description than does anthropology or history—less distorted, but not entirely free of distortion.

We have been examining conceptions of scientific claims and of scientific activity that are problems both for and in feminist theory. They are problems *for* feminist theorizing because they block the possibility of feminist transformations in the way scientists, philosophers, and social theorists think about science. They are problems *in* feminist theorizing because belief in at least traces of these dogmas hides from us the inadequacies in our understanding of how science is gendered.

GENDER: INDIVIDUAL, STRUCTURAL, SYMBOLIC—
AND ALWAYS ASYMMETRIC

Inadequate conceptualizations of gender are also a problem both for and in the feminist science critiques. The inadequacies within the critiques reflect in two ways the partial, and even perverse, understandings of gender that are characteristic of mainstream thinking. The first results from an excessive focus on just one or two of the forms in which gender appears in social life, obscuring the sometimes mutually supportive and sometimes oppositional but always important relationships in any given culture between the preferred expressions of gender symbolism, the way labor is divided by gender, and what counts as masculine and feminine identity and behavior. The second results from the faulty assumption that gender differences in individuals, in human

52

activities, and in symbolic systems are morally and politically symmetrical. In addition to the use of these two inadequate concepts of gender, there are also conflicting views about what strategies can best be used to eliminate androcentrism from knowledge-seeking. Let us consider these three problems in turn.

Some of the feminist science critics do not even recognize, let alone try to account for, the relationships between symbolic gender, the division of labor by gender, and individual gender. Since I pursue this issue in subsequent chapters, I will describe here just two examples of this kind of undertheorized approach to gender and science. In the first example, the issue is the support two forms of gender provide the third; and in the second, an opposition between two forms of gender motivates expressions of the third.

Equity studies focus on individual gender: on how women are discriminated against within the social structure of the scientific enterprise, and on the barriers the scientific enterprise and feminine gender socialization create for women entering and remaining in science. These studies explain the low representation of women in science courses, laboratories, scientific societies, and scientific publications in terms of these factors; and they criticize the characteristics of feminine identity and behavior encouraged by our culture that work against girls' and women's achievement of the motivation or skills to enter science. The proponents of equity recommend a variety of affirmative action strategies and resocialization practices for female children in order to increase the representation of women in science.

But these critics often fail to see that the division of labor by gender in the larger society and the gender symbolism in which science participates are equally responsible for the small number of women in science and for the fact that girls usually do not want to develop the skills and behaviors considered necessary for success in science. Until both the "emotional labor" and the "intellectual and manual labor" of housework and child care are perceived as desirable human activities for all men, the "intellectual and manual labor" of science and public life will not be perceived as potentially desirable activities for all women. The equity recommendations, moreover, ask women to exchange major aspects of their gender identity for the masculine version—without prescribing a similar "degendering" process for men. Feminists who have worked on these projects have exerted themselves heroically in the face of immense hostility for over a century, and I do not mean to trivialize their truly amazonian efforts. There certainly are good political reasons why they have not mounted a campaign to get men scientists involved in child care and in transforming their own gender needs and desires. But their efforts have not achieved the results they expected. One reason is that their shallow level of social analysis fails to locate those underlying causes of discrimination against women in science that are to be found in the gendered division of labor in social life and in science's enthusiastic participation in our culture's symbol-making.

In the second example, some of the "textual critiques" of science seem to imply that we could eliminate the androcentrism of science if only we would draw attention to the beliefs and behaviors commonly thought of as feminine but nevertheless characteristic of (men) scientists in history. They suggest that the growth of science has been promoted as much by intuitive thinking, by valuing relational complexes, and by nurturing attitudes toward both nature and new hypotheses as it has by formal logic and mathematics, by mechanistic views, and by the "severe testing" of hypotheses accomplished by "torturing nature." Thus they seem to say that challenging the symbolization of scientific activity as uniquely masculine could eliminate androcentrism from science.

Again, these critiques have proved valuable indeed; they have greatly advanced our understanding of how gender ideologies are used by science. But the recommendation ignores the conscious or unconscious motivations for such gender symbolizing provided by *conflicts* between divisions of labor by gender in the larger society and individual masculine identity needs. Gender totemism in science is often energized by perceived oppositions or conflicts between masculine identity needs and threatened or actual divisions of labor by gender.

The second inadequate conceptualization of gender involves the assumption that masculinity and femininity are simply partial but combinable expressions of human symbol systems, ways of dividing social labor, and individual identities and behaviors. Many feminist critics seem to say that it is possible to strip away the undesirable aspects of masculinity and femininity and thus arrive at attractive cores which, while partial, are morally and politically symmetrical. The problem for feminism, as these thinkers see it, is that science has confused the masculine with the human ideal when the human must also include the feminine. But femininity and masculinity are not so easily combined; central to the notion of masculinity is its rejection of everything that is defined by a culture as feminine and its legitimated control of

54

whatever counts as the feminine. Masculinity requires the conception of woman as "other," as Simone de Beauvoir pointed out.[18] Femininity is constructed to absorb everything defined as not masculine, and always to acquiesce in domination by the masculine. Thus this conception of gender difference cannot explain how in our culture, as in the vast majority of others, political power and moral value are monopolized by men at the expense of women. Gender is an *asymmetrical* category of human thought, social organization, and individual identity and behavior.

Finally, we can perceive very different assessments of gender in three proposals for the appropriate goal of a feminist critique of science. One approach argues that we should try to replace the masculine voice of science's past and present with a feminine voice. We should reverse the valuation of masculine and feminine interests in and ways of knowledge-seeking, leaving science differently gendered. We should want a science *for* women.[19] The second approach calls for the creation of knowledge-seeking not in the feminine but in the feminist voice.[20] This proposal holds that the exaltation of gender—masculine or feminine— is detrimental to a truly inclusive human science. The third approach claims that the goals of the first two are still limited by masculine metaphysical and epistemological frameworks. It urges that we try to eliminate the defensive androcentric urge to imagine a "transcendental ego" with a single voice that judges how close our knowledge claims approach the "one true story" of the way the world is. Instead, we should try to create "reciprocal selves" that are federated in solidarities—rather than united in essentialized and naturalized identities— and correspondingly "decentered" knowledge-seeking.[21] We should want a form and purpose for knowledge-seeking which, whatever their other advantages, would probably bear little resemblance to what we think of as science. In later chapters we will examine the tensions between these three proposals for the goal of a feminist criticism of science and the reasons why we should want to maintain rather than to eliminate these tensions.

An adequate theorization of gender will always lead us to ask ques-

---

[18]de Beauvoir (1953).

[19]This phrase is Dorothy Smith's (1977), though she may not have in mind the proposal described here.

[20]See, e.g., Hartsock (1983b).

[21]See, e.g., the discussions in *Signs* (1981); Marks and de Courtivron (1981); Flax (1984); Haraway (1985).

tions about the interactions between gender symbolism, the particular way in which social labor or activity is divided by gender, and what constitutes gendered identities and desires in any particular culture. These questions are pertinent to the culture of science in fifteenth- to seventeenth-century Europe as well as to the cultures that have supported science in later centuries. Furthermore, because of the "logical" asymmetry in the content and valuation of masculinity and femininity, it is a situation that requires explanation if we find men scientists carrying on what would appear to them to be characteristically feminine activity or holding the kinds of beliefs their culture identifies as feminine. We must ask questions about the often irrational relationship between the asymmetrical gender symbolism of activities and beliefs and the asymmentrical sexual order and forms of gendered personal identity. And we must critically examine the purposes and goals of the forms of knowledge-seeking envisioned as a result of the feminist revolution. To bring that revolution to the natural sciences requires that we deepen our understanding of the complexity of the relation between the different ways in which science is gendered, as well as that we more thoroughly abandon the dogmas of empiricism.

I have been arguing that scientific, philosophic, and popular understandings of natural science are particularly hostile to a feminist critique. This resistance may appear reasonable if one thinks of gender difference as either a "natural" elaboration of biological difference or as culturally created characteristics attributable only to individuals and their behaviors. And it will appear reasonable if one insists on an excessively empiricist understanding of "what science is."

A series of related dogmas of empiricism ground and provide justification for this hostility, securing an apparent immunity for the scientific enterprise from the kinds of critical and causal scrutiny that science recommends for all the other regularities of nature and social life. If we were to abandon these dogmas of empiricism, we could adopt the alternative view that science is a fully social activity—as social and as culturally specific as are religious, educational, economic, and family activities. We would then find valuable critical interpretive approaches to all the activities that count as scientific, as well as to those that make scientific activity possible: selecting problematics; formulating and evaluating hypotheses; designing and performing experiments; interpreting results; motivating, educating, and recruiting young people for the scientific work force; organizing that work force and the support services—in families and psychiatrists' offices, as well as in laboratories—that make it possible for some people to be scientists; selecting, funding, and developing the technologies necessary to carry out scientific inquiry and those that inquiry makes possible; assigning different social meanings and values to scientific reason and to moral, political, and emotional reason.

Feminism proposes that there are no contemporary humans who escape gendering; contrary to traditional belief, men do not. It argues that masculinity—far from being the ideal for members of our species—is at least as far from the paradigmatically admirable as it has claimed femininity to be. Feminism also asserts that gender is a fundamental category within which meaning and value are assigned to everything in the world, a way of organizing human social relations. If we regarded science as a totally social activity, we could begin to understand the myriad ways in which it, too, is structured by expressions of gender. All that stands between us and that project are inadequate theories of gender, the dogmas of empiricism, and a good deal of political struggle.

3 THE SOCIAL STRUCTURE OF SCIENCE: COMPLAINTS AND DISORDERS

Observers of the array of feminist criticisms of science have tried to rank them on a scale measuring how enthusiastically the scientific enterprise could itself acknowledge their legitimacy.[1] The criticism thought least threatening to science's self-understanding is that of unfair educational, employment, and status-assigning practices. Why is it, then, that after more than a century of attempts by women to enter science, the scientific work force today is so obviously gender-segregated? Why do patterns of vertical segregation still assign women primarily to low-status positions, and patterns of horizontal segregation designate certain areas of inquiry as women's and others as men's fields?

More pointedly, why is it that the scientific establishment has consistently resisted the education of women for careers in science, the employment of women in science, and the evaluation of women's work in science as equally deserving of public recognition and institutional support? Surely science's own rules require every fair-minded person to support the elimination of these kinds of unfair practices. And since eliminating them would not alter the nature and practice of science—or so many people believe—shouldn't such support be relatively easy to gather? Fair practice would add to the design and direction of scientific inquiry the skills and abilities of one-half of the human race; the "manpower" pool for science would be doubled. Both science's own self-corrective rules and obvious considerations of social justice would appear to require that the scientific enterprise acknowledge and

[1] See Keller (1982) for one of these "threat orderings."

respond positively to these criticisms. And since most people appear to believe that the changes called for would not threaten the epistemology or politics explicitly avowed by the scientific enterprise, why is there such a gap between women's expectations and the reality of science's response?

I do not intend to review here the extensive literature documenting the patterns of discrimination against women in science and speculating about the causes of these patterns. Instead, I wish to show that we must look at the mutually supportive relationship between individual gender, structural gender, and symbolic gender in order to understand the gap between science's self-image as a progressive, transcendentally valuable social enterprise and the reality of science today. The race, class, and cultural values of modern science could be similarly followed through the history of equity struggles. In tracing how gender values have shaped the scientific work force, we follow just one central strand through this tangled skein.

IS A WOMAN SCIENTIST A CONTRADICTION IN TERMS?

Let us look in some detail at one study that not only provides extensive quantitative and qualitative description of women's locations and achievements in science over a century but also highlights the gaps between the "progressive" rhetoric of scientism, the actual practices of both individual scientists and science as an institution, and the symbolic meanings of masculinity, femininity, and science. The broader social and political context in which discrimination against women in science occurs is part of gendered social relations more generally, and is also part of the psychic landscape within which individual masculine scientists think about themselves as well as about the nature of science.

Women's Struggles to Enter Science.
In her *Women Scientists in America*, Margaret Rossiter shows how women's struggles to enter science in the late nineteenth and early twentieth centuries occurred within two larger contexts that established the limits of their possible achievement.[2] "Women's historically subordinate 'place' in science (and thus their invisibility to even experienced historians of science) was not a coincidence and was not due to any lack of merit on their part; it was due to the camouflage intentionally placed over

---

[2]Rossiter (1982b). Subsequent page references appear in the text.

their presence in science in the late nineteenth century" (p. xv). Both genders worked out this arrangement as the result of the "partial convergence of two major, though essentially independent, trends in American history between about 1820 and 1920" (p. xv). One trend was evident in the rise of higher education and expanded employment opportunities for middle-class women. The other could be seen in "the growth, bureaucratization, and 'professionalization' " of American science and technology. The first trend permitted women to gain the kinds of science education earlier available only to men, and to get jobs within the scientific enterprise. The second trend ensured that the relationship between women's education, on the one hand, and their employment and prestige opportunities, on the other hand, would not be the relationship available to men—the one expected to be the norm for science.

> If success can be judged in numbers, women scientists had done very well indeed, for by 1940 there were thousands of such women working in a variety of fields and institutions, whereas sixty or seventy years earlier there were about ten at a few early women's colleges. This great growth, however, had occurred at the price of accepting a pattern of segregated employment and underrecognition, which, try as they might, most women could not escape. [p. xviii]

The increase in numbers by 1940 was the result of a century of heroic struggle. Through a variety of strategies the women's colleges were founded and began to offer science education to women. But the official justification for educating women was not that they could then obtain opportunities equal to those available to educated men, though this was in fact a goal of many women who supported and taught at the colleges as well as of many who entered their doors; rather, the public justification for women's colleges was that educated women could raise better sons. "Hardly anyone expected middle-class women to, or wanted them to, hold jobs outside the home—or to vote. Raising and teaching sons who would work and vote, however, were deemed to be such overwhelmingly important full-time tasks that it was felt that mothers must be educated through the secondary and, later, college levels" (p. xvi). Thus the opportunities available to educated women would be limited by familiar culturally created gender stereotypes.

> Even as women's educational level rose and their role outside the home expanded, they were seen as doing only a narrow range of "womanly"

activities, a stereotype that linked and limited them to soft, delicate, emotional, noncompetitive, and nurturing kinds of feelings and behavior. At the same time, the stereotype of "science" was seen rhetorically as almost the opposite: tough, rigorous, rational, impersonal, masculine, competitive, and unemotional. In terms, therefore, of nineteenth-century stereotypes of rhetorical idealizations, a woman scientist was a contradiction in terms. . . . Women scientists were thus caught between two almost mutually exclusive stereotypes: as scientists they were atypical women; as women they were unusual scientists. . . . Moreover, this conceptual element meant that much of the history of women in science would be worked out not simply in the realm of objective reality, of what specific women could or did do, but covertly, in the psychic land of images and sexual stereotypes, which had a logic all its own. [p. xv]

Another historian of the period points out that the increased public concern with discrimination against women in education and employment was by no means the result only of feminist ideas that women should have opportunities equal to those of men. One of the most compelling reasons to provide expanded educational and job opportunities for women was the phenomenon of the vast increase in Europe and the United States of "redundant" or "superfluous" women, as unmarried women were called in nineteenth-century England. In the United States, Civil War deaths eliminated three million married or marriageable men. From rural areas, villages, and small towns, young men in far higher numbers than women were drawn to new job opportunities in the large industrial centers and in the development of the West. In England, colonial expansion took some five million young people, mostly men, from Britain between 1830 and 1875. These vast imbalances in the sex ratio left millions of women, many the sole support of their children, with severely deteriorated opportunities for economic survival. How were these single women, educated only for the "occupation" of marriage, to support themselves?[3]

Our times are not the first in which poverty has been disproportionately a woman's lot. Social reformers of that day were aghast at women's situation, and the opening of higher education and careers in science to women was just one part of a more widespread campaign to provide economic options other than the marriages increasingly fewer women would be able to make. This demographic shift was itself one of the stimulants to the rise of the nineteenth-century women's

[3]Faderman (1981, 183–84).

movement: it was a response to—among other things—a real deterioration in women's economic opportunities.

Rossiter argues that the most interesting period of struggle for women occurred between 1880 and 1910. By 1880 the women's colleges were offering science education, and in greater numbers women began to try to obtain the laboratory apprenticeships, graduate degrees, memberships in prestigious scientific societies, and appointments and prizes that were available to similarly accomplished men. Yet even though this was a period of "great fluidity and innovation" in which "new roles and opportunities were unfolding at the same time that new persons were becoming available to fill them," women constantly found the expanding opportunities in science closed to them (p. xvi). Women's small successes in the 1870s and thereafter in joining scientific organizations and finding work in museums and observatories appeared to men as women's encroachment

> upon what had formerly been exclusively masculine territory. Such incursions brought on a crisis of impending feminization, and a series of skirmishes in the 1880's and 1890's resulted in the women's almost total ouster from major or even visible positions in science. Although still allowed to enter most areas of science, they could hold only subordinate, close to invisible, and specifically designated positions and memberships. [p. xvii]

Rossiter stresses that it took only a very few women scientists to raise in men's minds the threat of "impending feminization."

By 1910, "a new rigidity had set in," and "despite much protest by feminists of both sexes, women's subsequent experience in science was more one of containment within previously demarcated limits than expansion into newer and greater opportunities beyond them" (p. xvi). These limits took two forms. On the one hand, women could hold auxiliary and subservient positions in the scientific fields where men predominated: they could be scientific educators in high-schools or, at the low-paid and revolving-door levels of instructor and assistant professor, in colleges; they could be assistants or technicians in industrial or private laboratories; they could work as scientific editors. On the other hand, they could practice science in such new "women's" fields as home economics or "cosmetic chemistry." Thus "even though women could claim by 1920 that they had 'opened the doors' of science, it was quite clear that they would be limited to positions just inside the entryway" (p. xvii). Vertical segregation and horizontal segregation combined to ensure this result.

Lest we think women's struggles in science are over, studies of more recent periods reveal a continuation of the patterns Rossiter identified.[4] There is an occasional Nobel Prize winner, such as Barbara Mc-Clintock, and a few other extraordinary women scientists for whom public recognition still calls forth as much comment on their gender as on their scientific accomplishments. The hundreds of thousands of other women trained and working as scientists, are primarily to be found in the lower echelons of the scientific enterprise, and the achievements of the few who can find the resources to carry out independent research are systematically undervalued relative to similar achievements by men.

Contemporary observers of women in science make two interesting points about the logic of that "psychic land of images and sexual stereotypes" noted by Rossiter. In Michele Aldrich's review of studies of why fewer women than men choose to pursue science educations and careers, it becomes clear that the effects of gender stereotyping, which begin in the cradle, accumulate through childhood, adolescence, and adulthood in such a way as to systematically discourage women and encourage men to engage in the kinds of thinking and motor activity necessary for skill in scientific, mathematical, and engineering work. The literature shows how these kinds of thinking and motor activity are presented to children and adults as skills that men need in adult life—no matter what their occupation—in order to become and remain men, whereas for girls they are not only useless in adult life but detrimental to others' perception of them as feminine.[5]

While this literature does not directly address the issue of why science discriminates against women, it does suggest that the cultural stereotype of science that Rossiter described—tough, rigorous, rational, impersonal, competitive, and unemotional—is inextricably intertwined with issues of men's gender identities. It suggests that "scientific" and "masculine" are mutually reinforcing cultural constructs. Consequently, we should expect that in science more than any other occupation (except, perhaps, making war) it will take the presence of only a very few women to raise in men's minds the threat of feminization and thus of challenges to their own gender identity. The very

---

[4]See *Signs* (1978); Hornig (1979); Haas and Perucci (1984).
[5]Aldrich (1978).

existence of the gender order and of gender symbolism causally contribute to the low percentage of women scientists.

Other recent literature analyzes the reasons why women who meet admissions criteria to careers in science cannot seem to accumulate status in the ways their male colleagues can.[6] A man can succeed in reinvesting his prestigious education, publication record, and professional grants and appointments to create a capital of prestige; a woman's credentials apparently remain uninvestable, her prestige and status not accumulatable. The main reason for this difference appears to be that it is part of being a man to regard whatever women do as inferior, and to expect that other men (and women) share this evaluation. These studies show why scientific work known to have been done by women is invisible to men (and to many women) even when it is objectively indistinguishable from men's work. (One sociologist suggests that subconscious masculine resistance to citing a woman's scientific work may originate in the ancient but still viable belief that a man never mentions a "good woman's name" in public!)[7] What it means to be a man is, in part, to share in masculine control of women. Men's individual and collective needs to preserve and maintain a defensive gender identity appear as an obstacle to women's accumulating status within science. In other words, masculine gender identity is so fragile that it cannot afford to have women as equals to men in science.

Implications for the Social Studies of Science.
These and similar examinations of the systematic resistance of male scientists to women's equal participation in science raise a number of important challenges to traditional ways of understanding the history, sociology, and philosophy of science. In the first place, Rossiter argues that "at least part of the so-called 'professionalization' of science in the 1880's and 1890's begins to look more like a deliberate reaction, conscious or not, by men against the increasing feminization of American culture, including science, at the end of the century. Ejecting women in the name of 'higher standards' was one way to reassert strongly the male dominance over the burgeoning feminine presence" (p. xvii). It should be remembered, again, that it required only a small number of women to be perceived as a "burgeoning feminine presence" that

[6]Jonathan R. Cole, *Fair Science: Women in the Scientific Community* (New York: Free Press, 1979); Gaye Tuchman, "Discriminating Science" (review of Cole), *Social Policy* 11(no. 1) (1980); Rossiter (1982a, review of Cole and Hornig).
[7]Tuchman, "Discriminating Science."

threatened to bring about the feminization of science. Is the professionalization of work more generally a (largely successful) attempt to exploit the tension in American democracy between commitment to equal rights and commitment to a meritocratic understanding of the goals of democracy?[8] Appeals to efficiency, precedent, and preestablished or newly created standards appear most often when new groups are struggling to gain the equal protection by formal law or informal custom that was previously denied to them. Rossiter's point raises important questions for the sociology and history of work.

In the second place, Rossiter notes that the chronology of her story—before 1880, 1880 to 1910, and after 1910—"corresponds with few others in either American history, which relies heavily on such events as wars, economic depressions, and presidential administrations, or in the history of science, which has to date emphasized internal intellectual events such as scientific breakthroughs or 'revolutions' " (p. xvi). Lillian Faderman marks the period between 1880 and 1910–20 as one in which a radical shift occurred in the meanings and referents for the notion of proper womanly heterosexual behavior. It was during this period that the romantic friendships between adult and often married women, which had been idealized by both men and women for centuries, first began to be labeled deviant and pathological. Faderman argues that these friendships could be tolerated—even admired and extolled—by men as long as women had no real possibility of independent economic power or social status and as long as it was believed that sexual relations could not occur in the absence of a penis. She shows how the emerging nineteenth-century women's movement threatened to end the first condition, and the sexologists—especially Freud—made plausible the falsity of the second condition. The women's movement threatened vastly to increase the number of independent women, but the sexologists conveniently provided the science that could "prove" women's independence to be pathological. Faderman shows that the disappearance of enthusiastic, approving images of friendships between women from British and, later, American popular magazines and novels exactly correlates with the popularization of Freudianism first in England and, a decade later, in the United States.[9]

These two historical events alone—the entrance of women into science and the change in standards for appropriate womanly behavior—

[8]See Harding (1978; 1979).
[9]Faderman (1981).

suggest that some sort of radical shift in the history of sexuality and of social relations between the genders was occurring which cannot be captured by analytical categories based on men's understanding of their world. Rossiter's and Faderman's analyses confirm the more general claims of historians who began to try to tell the story of women by adding the history of women to traditional histories. They quickly discovered that the history of women can't be simply added, because traditional history's conceptual schemes do not permit women's natures or activities, or relations between the genders more generally, to be understood *as social* or, therefore, as historically significant. In particular, the periodization schemes of social and intellectual history (of which the history of science is a part) make invisible both women's activities and the effect these activities have on the "men's history" from which the biased periodization is derived.[10] Both American historians and historians of science appear, consciously or unconsciously, to have adopted conceptual categories whose systematic effect is to hide vast changes in real and threatened social relations between the genders and the effect of these changes on the ideas and practices espoused by Americans in general, as well as by the scientific enterprise. In order to understand the changes occurring in late nineteenth- and early twentieth-century science, evidently we need a fuller understanding than intellectual histories and histories of men's worlds can provide. We need to be able to see the events reported by these mainstream histories within histories of gender and sexuality. What would happen to our understanding of, say, the birth of modern science in the fifteenth to seventeenth centuries if we examined it *within* a history of social relations between the genders? How would we explain the rhetoric about science during that period which appeals both to visions of greater social justice for all and also to particularly violent expressions of misogynistic gender politics? (I return to this issue in later chapters.)

Finally, Rossiter's analysis challenges the accepted understanding of the social function of both the sociological and methodological norms of science. She points out that probably what kept women struggling to enter science in the face of obvious resistance was that they took at face value these sociological and methodological norms. They believed in "the optimistic liberal faith of the Progressive Era that an evil once documented . . . [would require] moral, well-behaved persons . . . [to]

---

[10]Kelly-Gadol (1976).

take corrective steps of their own accord" (p. 160). That is, they believed that the scientific enterprise intended to honor its expressed commitment to democratic, universalistic procedures for participation and advancement within the scientific community. And they believed in "the efficacy of demonstration. Once they had shown they could consistently equal or better the men in those laboratories to which they were admitted, their value would be evident and the barriers against them would fall everywhere" (p. 161). They took at face value science's claims that the only prerequisites for recognition and reward are scientific achievement.

Examining the history of women in science provides the clearest evidence that at least in the last century, we should look with suspicion at appeals to these sociological and methodological norms to justify the consistency of science with the democratic ethos. Consequently, should we not also take a skeptical attitude toward the legitimacy of science's ability to commandeer immense public resources? Whatever the functions of these norms, evidently they are not either a description of the reality of how science functions or a statement of ideals or goals for which most individual scientists or the scientific elite are willing to work. If women are systematically excluded from the design and management of science and their work devalued, then it appears that neither the assignment of status to persons within science nor the assessment of the value of the results of inquiry is, or is intended to be, value-neutral, objective, socially impartial. Instead, this discourse of value-neutrality, objectivity, social impartiality appears to serve projects of social control. An institution that insists it is already satisfying such goals, and can point to its rules for doing so, has created a powerful rhetorical device for legitimating its own biases and their adoption into equally biased law and public policy. Feminists are not the first critics of scientific ideology to raise this point, but our focus on the gap between rhetoric and practice with respect to such a socially distinctive and numerically huge class of citizens should make the point irresistible. It *should*—but, as Rossiter points out, the documentation of an evil rarely suffices to eliminate it.

Rossiter's and our reflections on the reasons why women have had to engage in such long and wearing struggles in their attempts to end discrimination in science suggest that affirmative action issues are not as easy to resolve as one might think. Already we have seen that even these supposedly least threatening feminist criticisms raise issues of the deleterious effect that the fragility of masculine identity and gender

67

symbolism have on the social structure of science and on the standards by which scientific achievements will be judged. They lead us to suspect that it is the conscious or unconscious intention of men in science to preserve this area of social activity for men only, especially when traditional forms of men's control of women are threatened. They make us aware of the conscious or unconscious hypocrisy of appealing to science's stated norms to defend scientific method and the actual social norms of science.

Perhaps most important for our thesis in this book, the discussion of affirmative action issues draws our attention to a curious coincidence: the emergence of severe threats to the existing gender order are often followed by new scientific definitions of women's inferiority and deviance. How much of the public enthusiasm that results in higher funding for scientific activities and greater prestige for scientists can be attributed to science's innovative ways of legitimating sexism, as well as classism, racism, and imperialism? The rise of IQ tests, behavioral conditioning, fetal research, transsexual operations, sociobiology, and many other scientific fashions can be observed with similar skepticism. What were the problems to which these scientific developments were responses? How did science benefit from its ability to define these problems in ways it said it could solve? What social conditions made its solutions to these problems plausible to other scientists and to the policy-makers who fund science? Can we possibly still imagine science to be in fact or principle value-free once we focus on the way the selection and definition of scientific problems escapes science's methodological controls? Can we ask these kinds of questions about the rise of the scientific world view itself?

From these perspectives it is clear that mere reforms of science cannot possibly resolve the equity issues. Instead, it appears that there will have to be revolutionary changes in social relations between the genders and in science's relationship to the societies that support it before it is no longer regarded as a contradiction in terms to be a woman scientist.[11]

ISOLATED GENIUSES OR INDUSTRIAL WORKERS?

There is an additional problem with the way the feminist criticisms of discrimination have been conceptualized. The ideology of science—the dogmas of empiricism—succeeds in directing our attention away

[11]See also the interesting discussion of this issue in Stehelin (1976).

from the facts about the social structure of science today. Most of us are brought up on an image of the scientific enterprise—how it works and what its goals are—that should be regarded as an extremely selective picture of pre-twentieth-century inquiry. As we will see, it mystifies our understanding even of seventeenth-century science and reflects virtually none of the socially interesting details of how the work of contemporary physics, chemistry, and biology is actually organized. It bears more resemblance to the military's recruitment literature than to a critical explanation of how scientific belief is produced. Thus the images of the production of scientific knowledge that some feminists have in mind when they raise the equity issues frequently do not reflect the realities of the social structure of science. Other feminists have a more realistic view.

Descriptions of the gap between the image of science used to recruit young people and the actual future awaiting scientists should be familiar by now. Thomas S. Kuhn, for one, pointed out that young persons must be recruited into science through implicit promises of heroic adventures on the frontiers of knowledge; they would not be enticed by learning that 99 percent of them will spend their lives merely solving the "normal science" puzzles that constitute the vast bulk of research today.[12] Nor, we can add, would they be enticed by the prospect of a "good job on the assembly line" in the production of scientific knowledge, which is the social form within which normal science is practiced.

The organization of the labor that produces scientific knowledge has changed historically, and it has changed in many of the same ways as the organization of the labor that produces such other goods as chairs and bread. Since the social relations of production processes affect the character of the products, we should not be surprised to discover that the scientific beliefs of different eras bear the distinctive marks of the social relations through which they were produced. The first chair any human constructed was no doubt different in kind from subsequent chairs produced through craft industry, and handmade chairs are different from factory-produced ones. Similarly, the first "handmade" scientific beliefs are different from the "factory-produced" ones that have predominated in the natural sciences at least since World War II.

[12]Kuhn (1970).

Changes in the Division of Labor by Class.
In contemporary depictions of medieval and early Renaissance medical education, the doctor characteristically reads to his academically gowned students from Aristotle while a barber or butcher dissects a cadaver beneath a barrier separating him from the doctor and students. Such pictures present clear messages about the expected activities and social status of physicians as opposed to those who actually came in physical contact with anatomy.[13] Other, equally advanced cultures with similarly strong social sanctions separating intellectual and manual labor have not—and probably could not have—produced experimental method.[14]

In contrast, as Jerome Ravetz has pointed out, seventeenth-century scientific knowledge was produced largely through craft labor. The emergence of this new social class whose members obtained the intellectual training required to conceptualize scientific experiments but were also willing to perform the manual labor required to execute these experiments appears to have been a necessary precondition for the emergence of scientific method.[15]

Since the nineteenth century, however, the production of scientific belief, like the production of other goods, has increasingly been organized along industrial lines. While chemistry had become industrialized by the late nineteenth century, it was not until after World War II that all the stabilized physical sciences achieved virtually complete industrialization, and many areas of social science research more recently have been transformed from craft to industrialized modes of research. (I say *stabilized* physical sciences, because new fields of scientific inquiry must initially be conceptualized and organized through craft labor.[16])

Thus the labor of producing scientific belief has been organized along the same rigid hierarchical lines as has the labor of producing furniture and breakfast cereals—or, for that matter, such services as health care. Managing the "factory" of science are government science policy advisors and the heads of research teams located in industry, in uni-

---

[13]See Stanley Joel Reiser, *Medicine and the Reign of Technology* (New York: Cambridge University Press, 1978), and the review in Harding (1978) for discussion of the obstacles to the development of diagnostic technologies created by social conceptions of the body.

[14]Zilsel (1942).

[15]Ravetz (1971).

[16]See Kuhn's discussion of this issue, and the first-person account of modern "craft labor" in Watson (1969).

versities, and in the government. These are the people who win Nobel Prizes, whose work is reported in scholarly journals as well as in *Time* magazine, whose writings philosophers and historians occasionally read, and whom most people have in mind when they think of a scientist.

Working closely with the managers of the scientific enterprise are the distributors of scientific knowledge. While pure research was once characteristically distinct from physical and social engineering and from applied science, the temporal gap between the two has steadily narrowed. As one commentator points out:

> Today, basic research is closely followed by those in a position to reap the benefits of its application—the government and the corporations. Only rich institutions have the resources and staff to keep abreast of current research and to mount the technology necessary for its application. As the attention paid by government and corporations to scientific research has increased, the amount of time required to apply it has decreased. In the last century, fifty years elapsed between Faraday's demonstration that an electric current could be generated by moving a magnet near a piece of wire and Edison's construction of the first central power station. Only seven years passed between the realization that the atomic bomb was theoretically possible and its detonation over Hiroshima and Nagasaki. The transistor went from invention to sales in a mere three years. More recently, research on lasers was barely completed when engineers began using it to design new weapons for the government and new long-distance transmission systems for the telephone company.[17]

Consequently, discovery and application, research and engineering, can no longer be distinguished; they have become part of the same process. In addition to the sheer capital required to conduct scientific research as well as to turn it into a socially distributable product, patent and copyright laws help ensure that this knowledge will be produced to benefit only those who also have the capital to distribute the results for profit or the power to organize and maintain policies of social control. Directors of the military, of the police, and of prison, health care, and mental health systems are examples of the latter.

Today, it is no longer possible to distinguish the individuals who manage the scientific enterprise from those who distribute its results. It is true that to the individuals involved, the research they do may often still appear distinct from its applications. But once we step back

---

[17]Zimmerman et al. (1980, 303–4).

from what individuals think about their own activities—what "the natives" think—to look at the overall structure of the production of scientific knowledge, such boundaries cannot be drawn so easily. As in many areas of human endeavor, the conscious goals of the individual actor often do not correlate positively with the explicit goals and implicit functions of the enterprise within which she or he works. The beliefs and behaviors of individual scientists provide an example of the culture-wide irrational belief and behavior whose description and explanation require the kinds of theories and methods of analysis found in some traditions of social science inquiry but in no traditions in the physical sciences.

The manager-distributors of science are only a small minority of scientific workers. One source estimates that "some 200–300 key decision-makers—primarily scientists—constitute the inner elite out of a total scientific work force of some two million."[18] Performing almost all of the labor actually required to produce scientific belief are the 1,999,700 or so technicians in laboratories and workers who manufacture the equipment and materials for scientific inquiry. (In the social sciences, these technical workers are the research assistants, interviewers, data gatherers and analyzers, computer programmers, and the like.) Finally, excluded from head counts of the scientific work force but crucial to the existence of science, there is the domestic staff—the vast numbers of clerical and plant maintenance people required to process the paperwork and day-to-day office functioning where research takes place, and to clean and repair equipment, offices, and laboratories. It would be reasonable to include in this domestic staff the armies of elementary, high school, and college teachers, guidance counselors, and science popularizers necessary to attract workers to careers in science and to train individuals for the various jobs. (Perhaps we should include all the socializers of infants as part of the work force responsible for the production of scientific knowledge, if it is true that masculine gender is an ideal precondition for becoming a director-manager of the scientific enterprise and feminine gender an ideal precondition for becoming a clerical worker or lab technician.)

Thus labor within the scientific enterprise is divided among three groups: the managers and distributors, the technical workers, and the domestic staffs. Only the first group conceptualizes and controls the execution of scientific research. But the social relations that produce

[18]Rose and Rose (1976, 33).

their selection and conceptualization of scientific problems are not limited to discourse and negotiation with one another, with their scientific traditions, and with "nature," as one would gather from the visions of science projected in science textbooks, histories, and philosophies. These social relations, and consequently science's picture of reality, are the product of the total social relations of the scientific enterprise, which are highly integrated with the larger social relations of the societies that support science. Individuals do not spring naked from the womb into the social relations of the laboratory table. Those social relations are but an extension of the social relations of all the other tables of the culture—in kitchens, schoolrooms, locker rooms, and board rooms.

The Integration of Social Relations.
How are the social relations of science integrated with the social relations of the larger society? Four aspects are particularly revealing.[19] Let us look in turn at science's preservation of absolute social status; at the division in science between the conception and execution of research; at the fit between the kinds of concepts favored in science and those necessary for "ruling"; and at the identity between the objects of inquiry and the objects of social policy.

First of all, the social hierarchy within science by and large preserves absolute social status: the social status scientific workers hold in the larger society. In science, we correctly imagine primarily white men of the upper classes when we think of scientists. We find women of all races and classes, men of color, and lower-class white men in far greater numbers when we look at precollege science teachers and laboratory technicians. The division of labor in science is consistent with the division of labor in the larger society, as a short walk through your local university or industrial laboratory will very quickly reveal. Those whom we think of as scientists, the science policy advisors and heads of research teams who make up less than 0.01 percent of scientific workers, are predominantly white and male and come from the upper middle-class backgrounds necessary to provide them with the motivation for, and funding of, the appropriate education. The higher ranks of technicians are predominantly white and include large numbers of women; these come primarily from middle-class backgrounds that can

---

[19]The analyses in this section and the next draw on Harding (1978), which in turn owes a debt to the critiques in David Kotelchuck, ed., *Prognosis Negative: Crisis in the Health Care System* (New York: Random House, 1976).

73

provide them with the undergraduate and graduate education and the skills necessary for supervising ongoing research activity. The lower ranks of technicians include far higher proportions of minority men and women and white women from lower middle-class backgrounds, who typically arrive with high school diplomas and often obtain some college education while employed. The clerical part of the domestic staff of science is almost entirely female, and plant maintenance workers in many areas of the country are disproportionately black or hispanic.

Such obvious social stratification may appear to some people to conflict with standard ways of understanding the point of industrializing labor. We are told that the industrialization of labor tends to destroy the uniqueness of individual labor that was characteristic of craft production—indeed, that one of its goals is to make all workers interchangeable parts of the industrial machine. The industrialization of labor is supposed to make irrelevant the unique social and natural characteristics and abilities of individuals; it standardizes labor routines so that individual workers possess no special knowledge of their laboring process. Bacon himself advanced this kind of goal for scientific method: "The course I propose for the discovery of sciences is such as leaves but little to the acuteness and strength of wits, but places all wits and understandings nearly on a level." He argued that "my way of discovering sciences goes far to level men's wits, and leaves but little to individual excellence; because it performs everything by surest rules and demonstrations."[20] If scientific method, and the introduction of scientific rationality into industry which the method justifies, "leaves but little to individual excellence," why does the division of labor in science preserve absolute racial, gender, and class status?

This question suggests one inadequacy in accounts that consider class the only analytically significant organizer of social relations—of theories that focus only on the complex history of struggles between the bourgeoisie and the proletariat (or their contemporary successors). This kind of analysis is invaluable for its ability to reveal many of the significant characteristics of the social relations of modern industrialized science, but it tends to obscure other important characteristics. Neither in the larger society nor in science does an analysis of the division of labor by class *alone* explain why it is that our "rulers" are predominantly white men, while women of all races and minority men are disproportionately represented in low-status jobs. Even in socialist

[20]Quoted by Van den Daele (1977, 34).

countries, whatever jobs are assigned high status (and these differ from country to country), it is primarily men, and men of the dominant racial and ethnic groups, who hold those jobs. We need to examine the divisions of labor by gender and race in order to explain the obvious social stratification in our lives. Once we realize that half of social labor is not to produce goods—commodities—but to reproduce people and social relations, it becomes clear that the division of labor by class cannot explain why there are so many white men in the top ranks of science and other high-status social enterprises and so few among nurses, social workers, secretaries, child care and domestic workers.[21]

The industrialization of labor has made workers interchangeable only within such other cultural categories as race, gender, and age. Furthermore, historians and economists have discovered ways in which the goals of men as workers are often in conflict with their goals as gendered persons or as whites in a racist, masculine-dominated society. Their class interests as workers do not always prevail in these conflicts.[22] Changes in the division of labor by class illuminate important elements of social relations. But considered apart from historical changes in the division of labor by gender and race, this kind of analysis can provide only a partial and distorted understanding of the social relations of science.

The preservation of absolute social status in the ranked division of labor inside the scientific workplace ensures that scientific workers will find it difficult to identify and organize around shared goals. The preservation of class, race, and gender status within science creates a reluctance to recognize shared goals and to organize across these divisions. Men of color and women who manage to rise to the top ranks of science may think that they share few work-related concerns with their sister and brother technicians—and the feelings of disaffiliation and distrust are frequently returned. Unions of scientific workers, like unions in other settings, have focused on improving salaries, benefits, and working conditions but not on redistributing the control of the scientific workplace to break down class, race, and gender stratification.

The second revealing aspect of contemporary scientific work is its reflection of a second major reason for industrializing labor: to separate the conception and execution of that labor, and to accumulate the conceptions and the knowledge of the execution in the minds and hands

[21]See Hartmann (1981b); Harding (1981).
[22]Hartmann (1981b).

of managers.[23] As a number of writers have pointed out, the execution of scientific research is now rarely done by the same persons who conceptualize that research, and even the knowledge of how to conduct the research is rarely possessed by those who actually do it: research is industrialized. Furthermore, given the preservation of absolute social status within the scientific work force, the conceptualizing of scientific problems remains the prerogative of white men. Adaptations of Taylorism into the physical and social scientific enterprises were crucial in moving the production of scientific knowledge from craft to industrial models, and in accumulating the knowledge of how to conduct research in the hands of the managers. Unions are the only organized resistance to Taylorism. But again, because unions focus on wages and benefits rather than on increasing workers' control of the laboring process, because unions are also bastions of white and masculine power, and because working-class interests as defined by unions have rarely included antiracist and antisexist interests, we should expect that the accumulation of racist and sexist power within science is little obstructed by unions. Indeed, class-based criticisms of the mode of production inside and outside science have rarely raised fundamental race and gender issues.

Third, the conceptualizing of the social and natural world is part of the labor of "ruling," and modes of ruling and codes for understanding nature and social life fit together and need each other.[24] In the physical sciences, conceptions of nature as passive but threatening to human life, and as resistant to inquiry, legitimate aggressive and defensively justified manipulations of nature and social life. These manipulations increase economic productivity and political power that benefit only the few; indeed, the very definition of many scientific problems in terms of ignorance about how to technically manipulate nature—though they often are fundamentally political and moral problems—reserves expertise for ruling groups. Consider, for example, the groups (in good health) that benefit from defining the problem as finding a cure for cancer vs. those that benefit from defining the problem as eliminating the causes of cancer. In the social sciences, conceptions of humans as passive recipients of external stimuli, and of social groups either as determined by the natures of their members and their environments (naturalism) or as systems of equally arbitrary customs, rules, and

[23]Braverman (1974). See also Sohn-Rethel (1978); Hartsock (1983b; 1984).
[24]Smith (1974; 1977; 1979; 1981) has discussed this issue with respect to sociology; her arguments are generalizable to the natural sciences.

meanings (intentionalism), intentionally or unintentionally legitimate the exercise of the social controls required to increase productivity while accumulating profit and control in the hands of only a few.

Fourth, it is no accident that in both the natural and social sciences, the objects of inquiry are the very same objects that are manipulated through social policy. It is not that the results of scientific research are misused or misapplied by politicians, as the ideology of pure vs. applied science holds. Rather, social policy agendas and the conceptualization of what is significant among scientific problems are so intertwined from the start that the values and agendas important to social policy pass—unobstructed by any merely methodological controls—right through the scientific process to emerge intact in the results of research as implicit and explicit policy recommendations. Suppose the problem is conceptualized as one of population control, and the reproductive practices of poor and Third World women are defined as the location of the problem. If changing these reproductive practices is considered a technological rather than a political problem, then the results of research will recommend abortion, sterilization, and the distribution of contraceptive pills for poor and Third World women. How could the results of research be any different?

And the results of this inquiry research are "inscribed" with these racist, classist, and sexist social policies in spite of the availability of alternative information about these social issues. For example, it is widely known that unequal consumption of natural resources by the rich and the poor actually makes high-consuming classes and cultures the cause of the low standard of living that population control for low-consuming classes and cultures is proposed to resolve. In the social sciences, race research has consistently formulated its problem as to determine the characteristics of Blacks and of Black social relations that are responsible for the low social status of Blacks, rather than those of racist institutions and white social practices. Traditional gender-role research has formulated the problem as lack of success by girls and women, rather than the obstacles that masculine-dominated social institutions raise to women's success and the excessively narrow conception of success that men characteristically hold. Industrial management and so-called "human relations" formulate labor problems as how managers can better control workers and make them happier with less power, rather than how to restructure work along more democratic lines.

These four aspects of the social relations of science demonstrate that

77

classist, racist, and sexist social relations are as central to the organization of science as they are to the organization of social life more generally. This integration permits the conception and execution of labor to be separated with relatively little resistance from workers throughout science's stratification. Because this separation feels natural to everyone involved, it permits a coherence between the scientific conceptualizations of nature and inquiry and the concepts useful for ruling in societies organized in class, race, and gender hierarchies.

Tensions and Contradictions.
Nevertheless, the integration of social relations is not perfect, and these places of incomplete integration provide the origins of valuable tensions and contradictions within the scientific enterprise.

First of all, the degree of integration varies among levels of scientific workers. It is highest for white men of the professional class; they hold the same status within science as they do in their families, in daily social life, and in their cultural mythologies. For minority and poor men, however, who often have significant status in their communities and families if not in the dominant culture or on the job, the integration is less than perfect. For women of every class and race, the integration is high—they have low status at work and their domestic labor is itself of low status. But the double-day of work that is the condition of their presence in the wage-labor force reveals most clearly the real social relations that maintain the status of the managers of the scientific enterprise.

Furthermore, everyone is aware that in spite of the vast differences in status, scientific workers share a certain degree of functional inter-dependence. The director of a project may get the credit, but the production of scientific knowledge requires the coordinated labor of workers at all levels. The career of a director who has had eight years or more of graduate and postgraduate training, who earns more than $100,000 per year, and who sits on national policy boards can be ruined by the error of a graduate student, technician, or plant maintenance person—as we can see in cases of faked research by protégés, the Three Mile Island nuclear disaster, and the mishaps that have befallen spacecraft.

In the second place, there is a tension between the ethic of science and the reality that scientific workers observe. The potential social value of increased knowledge appears to be immediately and unquestionably obvious, and it is the origin of the ethic that makes scientific

78

research a good in itself. The factory worker may wonder about the value of a new flavor of cat food or the nineteenth brand of can opener she is involved in producing, but the potential value of a cure for cancer or an alternative energy source is immediately apparent to scientific workers as well as to the general public. Critics often question the priorities of basic research and the uses to which the knowledge is put, but never the assumption that knowledge provides power to improve the conditions of human life. Young scientists are recruited through appeals to the ethic of scientific inquiry.

However, as we noted earlier, few of those so recruited will be able to do the kind of pathbreaking research that will earn them Nobel Prizes and secure niches in the history of science; the vast majority will become assembly-line technicians. Furthermore, for women and for men of color, because the priorities conceptualized by white males often create ambivalences about the social value of particular projects, research priorities may differ from those of their private lives outside science. Who would *choose* a career goal of building bombs, torturing animals, or manufacturing machines that will put one's sisters and brothers out of work? Thus a tension is increasingly obvious between the ethic that draws young people into the arduous training necessary for a career in science and the realities produced by the actual projects for which they are recruited and the sheer numbers of scientifically trained workers.

Clearly, there is a damaging gap between the assumptions about scientific inquiry that ground popular images and the assumptions required to explain how the results of research both are and should be produced within the actual social structure of contemporary science. The image tells us about a single individual, beholden to no social commitments but only to the search for truth, who creatively identifies and conceptualizes problems worthy of inquiry, invents methods of asking nature questions, and achieves clear and value-neutral results. The reality of industrialized scientific production requires a set of concepts that can capture the relations between different social divisions of labor and the inquiry products they produce—between race, class, and gendered divisions of labor and the form and content of the scientific claims produced through this labor.

For over a century women have struggled to enter science as equals to men. One account of the turn-of-the-century period of this struggle suggests that the professionalization of science may itself have been a device to preserve the direction of scientific inquiry for elite, white men. Moreover, standard ways of periodizing history preclude analyses of how real and threatened changes in social relations between the genders have affected the history of science. Furthermore, the explicit sociological and methodological norms of science in fact function, at best, as rules for how to treat the work and activities of white male scientists. Finally, the status of science and the prestige of scientists appear to have benefited from science's ability to provide those kinds of definitions of scientific problems and their solutions that support masculine dominance. Was it entirely a coincidence that sexology began to gain status as a science hot on the heels of the nineteenth-century women's movement and women's agitation to enter science?

In examining the actual social structure of contemporary physical science, we can see that the image of scientific activity projected by philosophers, historians, and other science enthusiasts does not reflect the normal way scientific belief is produced today. The men to whom women want to be equal are the directors of the scientific enterprise—a tiny proportion of those whose work is required to produce scientific belief—and a condition for holding such positions is the implicit acceptance of science's acquiescence and support of the sexist, racist, and classist organization of labor and social status in the general society.

These conclusions are certainly not politically or spiritually uplifting. On the other hand, exactly because science's social hierarchy so closely mirrors the social order "outside," any progressive changes that can be brought about in the social structure of science should have rapidly escalating consequences for the larger social order. After all, though naiveté is to be recommended—at most—only for the young, Rossiter drew our attention to the fact that the naiveté of nineteenth-century feminists played an important role in making possible the twentieth-century women's movement, with all the changes in social life to which it has contributed. And we shall see that some women scientists have been able to locate themselves in the social structure of science in ways that have produced far-reaching emancipatory consequences.

This chapter's focus on the actual social structure of contemporary science is intended to introduce a dose of reality into the fanciful and dangerous picture of the isolated genius that is commonly presented by mainstream history and philosophy of science. And it is intended to alert us to the necessity of understanding gender not just as a char-

80

acteristic of individuals and their behaviors, and not just as a way of organizing social meanings—as gender totemism. We must also look at how these forms of the gender order shape and are shaped by the actual divisions of labor by gender, class, and race.

4 ANDROCENTRISM IN BIOLOGY AND SOCIAL SCIENCE

In the last chapter, we saw that the feminist challenges often considered least threatening to science—the equity issues—lead to the possibility that equal opportunity for women in science requires a radical reduction in gender stereotyping, in the division of labor by gender, and in the defensive fragility of masculine identity. It may even require the complete elimination of sexism, classism, and racism in the societies that produce science. These are hardly mere reforms in social relations.

Regarded as somewhat more threatening than the affirmative action challenges is the contention that masculine bias is evident in both the definition of what counts as a scientific problem and in the concepts, theories, methods, and interpretations of research. This charge has been made against both the social sciences and biology, but physical scientists and their philosophical interpreters—who think there is little or nothing they can learn from the social and life scientists—tend to believe that such feminist criticisms have no relevance to the physical sciences. Therefore, the feminist charge of masculine bias, while more threatening to science-as-usual than the equity challenges, still appears to most scientists—feminist or not—to leave untouched (and untouchable) physics, chemistry, and the scientific world view. In Chapter 2 we saw that this faith in the inherent immunity to social influences of physics, mathematics, and logic is unjustified. Before examining the feminist criticisms of the social and life sciences, let us clarify their relevance to our understanding of all the physical sciences.

82

ARE SOCIAL SCIENCE FINDINGS IRRELEVANT TO THE CONDUCT OF NATURAL INQUIRY?

One side of a long history of argument in the philosophy of social science claims that the value-laden character of the social sciences has three origins, each of which makes it inadvisable to model social inquiry on physics. For these philosophers, the philosophy of the natural sciences is regarded as irrelevant to the philosophy of the social sciences. But these philosophers appear to agree with their opponents that social inquiry is irrelevant to natural inquiry.[1] That claim requires separate arguments, which neither side ever provided.

Both the "naturalists" and their opponents, the "intentionalists"— as the two parties to this dispute have come to be named—agree that the social sciences have a kind of subject matter different from that of the natural sciences: the former deal with humans and cultures which, in contrast to inanimate matter, constitute themselves through significances, meanings, and histories. Unfortunately, the naturalists argue, the social meanings and values characteristic of this subject matter all too often seep into the results of inquiry. Nevertheless, the naturalists insist that these social phenomena can be explained in the same kinds of causal terms as can purely physical phenomena, and that stricter adherence to the methodological controls so effective in physics will successfully eliminate social values from social inquiry. There is just one scientific metaphysics and one scientific methodology: the ones characteristic of physics.

The intentionalists reply that what is unfortunate in social inquiry is just the tendency to impose this kind of alien, physicalist, conceptual scheme on humans' understandings of their own cultures and activities. Instead, they say, the inquirer must draw on, activate, his/her own complex of social meanings and values in order even to distinguish social from natural events and processes. How would we know we were observing a salute to the flag rather than a muscle reflex without imputing social meaning to some events and not to others? And it is the "natives' " social meanings that are important, not the inquirer's, if we would avoid ethnocentric distortion in the accounts of what we observe.

[1] See Fay and Moon (1977) for a review of traditional perceptions of the problems with each side of this dispute (they do not identify the problem I am raising here, however).

In the second place, the naturalists continue, explanations of social life must account for more variables than do explanations of natural phenomena. Social inquiry is just harder than natural inquiry. Third, the social sciences are younger and less mature than the natural sciences; in time, they will move from preparadigmatic fact collection and disputes over assumptions to "normal science" agreement about theoretical assumptions, methodological constraints, and programs of research. But the intentionalists also dispute these purported origins of social science's value-ladenness.

Both sides to this dispute assume that the social sciences' problems of maximizing objectivity and value-neutrality have no parallels in the natural sciences. But there are several reasons to find this assumption implausible. In the first place, the social sciences have tried to imitate the dispassionate, objective methods supposedly characteristic of physics. Even the minority voices of the *verstehen*, "humanist," and hermeneutical approaches (the main tendencies within intentionalism) to social inquiry still value the objectivity and empirical fit between theories and observations that are seen as the strength of the natural sciences, believing that different kinds of methods and a different ontology will best screen out distorting infusions of the inquirer's values into the results of social research. But we can still reasonably ask whether the social biases of the social sciences are only a result of their differences from the natural sciences. Rather, do they not perhaps reveal a fundamental gap between the explicit epistemology and prescriptive methodologies of the natural sciences and the actual processes through which any inquiry—natural or social—has occurred or must occur? Real as the problems mentioned above may be, perhaps they are insufficient to account for all of the value-ladenness regarded as objectionable in social inquiry.

More important, as we discussed earlier, natural science is a social phenomenon. It has been created, developed, and given social significance at particular moments in history in particular cultures. Many of the claims made by feminist critics about how white, modern Western men of the administrative/managerial class tend to conceptualize social phenomena can be directly applied to the story of natural science as it is handed down in the history and philosophy of science, in science texts, and by the "greats" of modern science. If gender is a variable in the most formal structures of beliefs about the boundaries between nature and culture, or the fundamental constituents of socially constructed realities, why should we assume that the formal structures of

84

natural science belief are immune? Inadequacies in the choice and definitions of problematics and in the design and conduct of research in the social sciences reappear in the partial and distorted self-consciousness of the philosophy of both social and natural science, as well as in the favored accounts of the history and social structure of science. The social practice of and beliefs about natural science are appropriate subjects for social inquiry, but we need degendered social sciences and philosophies of social science to provide objective understandings and explanations. What is the point of a philosophy of science that cannot account for the obvious historical successes and limitations of the enterprise it would explain and thereby direct? If economic, political, psychological, and social regularities are significant in creating those historical successes as well as sometimes in blocking "the growth of knowledge," then gender in its threefold expressions—the totality of social relations between the genders—will also be found at the center of those regularities.

For these two reasons, the feminist criticisms of bias in social sciences have relevance far beyond their explicit subject matter; they have relevance to our analysis of all science.

FIVE SOURCES OF ANDROCENTRISM IN SOCIAL INQUIRY

In their introduction to *Another Voice: Feminist Perspectives on Social Life and Social Science*, an early collection of feminist criticisms of the social sciences, Marcia Millman and Rosabeth Moss Kanter identify six problematic assumptions that have directed sociological research.[2] Because these assumptions appear in other social sciences as well, we can use five of their six categories to grasp the depth and extent of the feminist charge that masculine bias in social inquiry has consistently made women's lives invisible, that it has distorted our understanding of women's and men's interactions and beliefs and the social structures within which such behaviors and beliefs occur. (The sixth assumption concerns the goals of social inquiry, an issue I shall take up later.) It is useful to focus here on an early set of feminist criticisms of the social sciences as a basis for reviewing what is generally accepted by feminist scholars today. The Millman and Kanter analyses have been elaborated and refined, but these scholars of the 1970s identified problems that have remained crucial areas of feminist concern.

[2]Millman and Kanter (1975). Subsequent page references to this collection appear in the text.

First, they point out that "important areas of social inquiry have been overlooked because of the use of certain conventional field-defining models" (p. ix). For example, the role of emotion in social life and social structure tends to become invisible in sociological analyses that focus exclusively on the role of Weberian rationality. Sociological images of the social actor tend to feature only two types of humans, for neither of whom are self-consciousness of feeling and emotions a crucial element in beliefs and behaviors: either the "conscious, cognitive actor . . . consciously wanting something (e.g., money or status) and consciously calculating the merit of various means toward an end," or the "unconscious, emotional actor . . . 'driven' or 'prompted' by a limited number of 'instincts,' 'impulses,' or 'needs' to achieve, affiliate, or do any number of things that merely surface as ends or means."[3] In neither case is awareness of feeling or emotion seen as significant in the reasons for or causes of people's actions and beliefs, or as an element of social structure, and yet such consciousness of feeling appears to be an obvious and important element in our own and others' beliefs and behaviors. We can wonder if this tendency to ignore the social role of conscious emotion is exacerbated by the combination of a cultural stereotype and a second sociological assumption. On the one hand, gender stereotypes present only women as motivated by conscious feelings and emotions; men are supposed to be motivated by calculation of instrumental or other "rational" considerations. On the other hand, social science assumes that it is primarily men's activities and beliefs that create social structure. Are not both men and women often motivated to adopt beliefs and behaviors, to support policies and institutions, by an awareness of their own feelings of love, affinity, anger, or repugnance?

Second, "sociology has focused on public, official, visible, and/or dramatic role players and definitions of the situation; yet unofficial, supportive, less dramatic, private, and invisible spheres of social life and organization may be equally important" (p. x). Such restrictive notions of the field of social action can distort our understanding of social life. For instance, they tend to make invisible the ways in which women have gained informal power. They hide the informal systems of men's sponsorship and patronage, that both ensure coveted career paths for professional men and isolate women employees—thereby

[3] Arlie Hochschild, "The Sociology of Feeling and Emotion: Selected Possibilities," in Millman and Kanter (1975, 281).

circumventing the overt goals of affirmative action programs. They obscure the ways in which the accomplishments of "geniuses" in the history of art, literature, politics, and the sciences have been made possible only through an analytically invisible substructure of women's support systems and social networks (p. 33). They make invisible the role of social interactions in local settings in community life—the settings where women predominate—in shaping those communitywide interactions and policies where men appear as the creators of social structure (p. xii).

Third, "sociology often assumes a 'single society' with respect to men and women, in which generalizations can be made about all participants, yet men and women may actually inhabit different social worlds" and this difference is not taken into account (p. xiii). Jessie Bernard has argued, for example, that the same marriage may constitute two different realities for the husband and the wife; this fact invalidates generalizations about marriage and family life that do not identify and account for the differences in position and interests.[4] Similarly, economist Heidi Hartmann points to the "battle between the genders" within the family over housework, which is responsible for giving women and men different interests in a wide array of public policy issues.[5] Additional analyses reveal many other kinds of interactions and institutions where women more then men are forced to lower expectations and rationalize discomfort in order to gain economic or social/political benefits.

The single-society issue in sociology is related to conceptual problems in the social sciences noted by other feminists. The common assumption that a particular social structure or kind of behavior is functional for the agents or the society usually ignores the misfit between women's consciousness, desires, and needs and the roles assigned to women.[6] Beyond and across adjustments to race and class hierarchies, women are forced to accommodate their natures and activities to restrictions they have not chosen. The gap between their consciousness and the expected behaviors they exhibit is what has made the consciousness-raising achievements of the women's movement such an important scientific as well as political resource. Male-dominated social

[4]Jessie Bernard, "The Myth of the Happy Marriage," in Vivian Gornick and Barbara K. Moran, *Woman in Sexist Society: Studies in Power and Powerlessness* (New York: Basic Books, 1971).

[5]Hartmann (1981a).

[6]See Westkott (1979).

orders are not functional for women, but one cannot easily detect that fact simply by observing women's behaviors.

Equally problematic implications arise from the suggestion in anthropology that the models of social structure—indeed of the very boundaries of the social—assumed by men in all cultures appear to be peculiarly consistent with the anthropological models of Western masculine investigators.[7] Social actors who are women appear to make significantly different and broader assumptions about what constitutes social interaction and social structure than do either the men in their own culture or (masculine) social scientists. Pertinent to our interests is the apparent fact that much of what men count as nature—as outside of culture—is part of culture for women.

Sociologist Dorothy Smith has analyzed the fit between, on the one hand, administrative models of social structure and the administrative personalities and interests to which men of all classes in our culture aspire, and, on the other hand, the conceptual structure of sociology.[8] She argues that the conceptual apparatus of sociology is part of the conceptual apparatus of ruling in societies with our kind of primarily masculine and administrative "rulers." For instance, she points out that the sociological category "housework" has been made part of a conceptual scheme wherein all human activity is either work or leisure, a dichotomy that more accurately describes men's lives than women's. Child-raising, cooking, house care, and the like are certainly both work in the sense of socially useful labor *and* leisure in the sense of oft-chosen and pleasurable activity, but for women they are both more and less than these categories can capture. Child-care in particular seems distorted by this dichotomy. It is less than babysitting, which has the fixed hours, limited responsibilities, and economic return (albeit a low one) of wage labor. But its value to women and, indeed, to society is far more than that of a bridge game, a trip to the beach, or most kinds of wage labor.

Moreover, in industrialized societies, it is convenient for the administrator-rulers to divide all human activities into time spent at work for others for pay and time spent at leisure, which it is the individual's responsibility to organize and maintain. Since leisure is regarded as a matter of private, individual choice, only labor for others—at best—requires social support. Welfare-state capitalism has had to accom-

---

[7] See Ardener (1972); Smith (1974).
[8] Smith (1974; 1977; 1979; 1981).

modate itself to increasing demands for public support of women, children, the aged, sick, and unemployed, yet policy-makers and analysts still tend to see these as merely social programs in contrast to the truly political programs directing wage-labor and foreign policy. Smith argues that sociology's replication of the conceptual categories of industrial capitalism makes sociology part of the ruling of our kind of society. (We might further ask whether Marxism's tendency to insist that the fundamental locus of politics is in the economic world—narrowly construed as the world of production—does not also replicate and thus support industrial capitalism's conceptual world.[9]) Smith's arguments appear applicable to many of the conceptual frameworks of other social sciences as well. Far from inhabiting a single society, women and men appear to live in different worlds. But it is only the men's world that social science takes to be the social world.

Fourth, "in several fields of study, sex is not taken into account as a factor in behavior, yet sex may be among the most important explanatory variables" (p. xiv; from the theoretical perspective of my study, it is gender difference, not sex difference, with which Millman and Kanter are concerned). There is, for example, the failure to analyze the impact of the gender of the classroom teacher or the physician on the interactions these people have with girls and boys, women and men, and the failure to examine the effect of stereotypical masculine models of the artist, the scientist, or the successful person on women's motivation to enter traditional masculine fields and to be recognized as successful in them.

As my parenthetical remark in the preceding paragraph suggests, confusingly but intimately intertwined with discussions of sex as a variable in social action are the issues of gender as a variable in history and in contemporary social life. Historian Joan Kelly-Gadol indicates that feminist scholars in history, too, have shown that the "sex" of social actors has been ignored as an explanatory variable, even though it is probably the single most significant variable in history. Her point is not that biological differences between the sexes have primarily determined the course of history; rather, she is elaborating Simone de Beauvoir's claim that "woman is made, not born." Social constructions of sexuality and gender have been responsible for assigning women and men to different roles in social life. Thus men, too, are "made, not born," and they are also distinctively men in the gender-specific

---

[9]Balbus (1982) is one critic who makes this point.

sense—not accurately presented as representative of "humanity." Kelly-Gadol argues that history has been shaped not only by distinctively masculine needs and desires but also by the socially constructed activities of women; thus studies assuming that women's natures and activities are fundamentally biologically determined and that men's socially created natures and activities are entirely responsible for social patterns doubly distort women, men, and social life.[10] Millman and Kanter point out, "When male sociologists (or men in general) look at a meeting of a board of trustees and see only men, they think they are observing a sexually neutral or sexless world rather than a masculine world" (p. xiv). If we substitute "genderless" for "sexless," we can see that the problem these critics are addressing is that only women are assumed to be the bearers of gender and only men the bearers of culture.

Fifth, "certain methodologies (frequently quantitative) and research situations (such as having male social scientists studying worlds involving women) may systematically prevent the elicitation of certain kinds of information, yet this undiscovered information may be the most important for explaining the phenomenon being studied" (p. xv). Criticism of an excessive preference for quantitative measures certainly does not originate with feminists. What is new in the feminist criticisms is the suspicion mentioned earlier that the preference for dealing with variables rather than persons "may be associated with an unpleasantly exaggerated masculine style of control and manipulation" (p. xvi).

The impact of the gender of the researcher on the adequacy of the results of inquiry has several dimensions. There is the obvious problem that for social reasons men do not have real access to many women-centered aspects of social life, either in our society or in other cultures. Such indirect access as they gain is primarily through masculine informants whose knowledge of women's activities is both limited and shaped by local ideological beliefs; if they do gain direct access, their presence changes the situation they are observing or the responses they elicit beyond the changes expected in interview or participant-observer situations. In part, this series of methodological problems explains the excessive focus in social science on the official, visible, and/or dramatic performers and social situations, for it is primarily these actors and this world to which (masculine) observers have access, and it is these

[10]Kelly-Gadol (1976).

actors and this world that masculine informants think most important in the cultures studied.

The historical dimensions of this problem are the subject of constant comment in anthropology, for the classic ethnographies were primarily collected by men who had either little or only distorted access to "native" women informants and to women's activities.[11] Thus the existing reports of what women actually believe and do now or at any other time in history must be regarded as far less reliable than the reports of men's beliefs and activities. The latter are also questionable, however: men are as gendered as are women, and everyone knows that men report to each other different aspects of their beliefs, desires, and behaviors than they report to women. Selective and distorted communication therefore occurs *between* men as well as between men and women. All these methodological limitations raise again the question of the suspicious fit between the concepts and theories favored by social science and those favored by men in every culture.

The five foregoing highlights of feminist criticisms do not pretend to provide a complete list of the ways in which it is clear that distinctively masculine bias has permeated the social sciences. There are more problems in sociology than the present brief account can address; in psychology, anthropology, history, and economics as well, biases peculiar to the subject matters and methodologies of each field similarly distort understandings of the social order.[12] But this outline is sufficient to indicate that feminist criticisms do severely challenge social science's self-perceived attempts to be value-neutral, objective, and dispassionate. As I have already suggested, it is not at all clear that these problems are solely the consequences of social science's different subject matter, variable complexity, and immaturity relative to the natural sciences.

More important to this study is that all these problems reappear in the favored philosophies, histories, and sociologies of natural science— in the *social* studies of science, as well as in popular understandings of science. Important areas of the social aspects of natural science, too, "have been overlooked because of conventional field-defining models." Traditional social studies of natural science, too, have focused on the "public, official, visible, and/or dramatic" at the expense of perhaps equally important "unofficial, supportive, less dramatic, private, and invisible spheres of social life and organization." The social studies of

[11]See, e.g., Leacock's (1982) discussion of this issue.
[12]For more extensive analyses of feminist criticisms of the social sciences, see Andersen (1983); Bernard (1981); and the frequent review essays in *Signs* (1975 *et. seq.*).

science, too, often assume "a 'single society' in which generalizations can be made about all participants, yet men and women may actually inhabit different social worlds" in the natural sciences. In the social studies of science, too, gender "is not taken into account as a factor in behavior yet may be among the most important explanatory variables." Finally, methodologies and research situations in the natural sciences, too, "may systematically prevent the elicitation of certain kinds of information" that "may be the most important for explaining the phenomenon being studied."

I have argued that contrary to the dogmas of empiricism, the same kinds of analytical categories are appropriate for understanding science and society, and that science is not just a particular set of sentences or a unique method but a comprehensive set of meaningful social practices. If self-understandings of the nature and purposes of science shape the practices of science, then—contrary to empiricist dogma—the kinds of beliefs physics and chemistry tend to produce should be explained in the same ways that we explain the kinds of beliefs produced through anthropological, sociological, psychological, economic, political, and historical inquiry.

VULNERABLE POINTS IN BIOLOGICAL INQUIRY

Biology is thought to be at least in principle less subject to the social passions that wash across the fabric of social inquiry. It would be possible to match the preceding list of biased assumptions in the social sciences with a list of those widely found in biology.[13] However, my intention in this chapter is not to perform a thematic survey of the vast literatures identifying masculine bias but instead to make vivid that such bias does occur and to stimulate thinking about the causes of and solutions to such bias. Therefore, my strategy here is to look at one illuminating analysis of the points at which biological inquiry is vulnerable to masculine bias and, at the same time critically reflect on the assumptions about gender and science that guide this analysis.

Biologists argue that in two kinds of studies of biological sex differences, evolutionary studies and neuroendocrinological studies, the results of research intersect in such a way as to make a powerful case

[13]For a representative sample of such criticisms, see Bleier (1984); Brighton Women and Science Group (1980); Gross and Averill (1983); Haraway (1978); Hubbard, Henifin, and Fried (1982); Hubbard (1979); Leibowitz (1983); Lowe and Hubbard (1983); Sayers (1982); Tobach and Rosoff (1978; 1979; 1981; 1984).

92

for biologically determined sex roles. These studies, individually and in their purported joint implications, have been the particular target of feminist concern in biology. If the findings of these studies were indeed plausible, it would be even more difficult than it now is to argue that moral considerations and enlightened public policy should lead to the end of masculine dominance and the restriction of women's opportunities; indeed, were these biological determinist arguments true, a "woman scientist" *should* be a contradiction in terms. (Later I will agree with the biological determinists that there is a contradiction, but draw different conclusions about it.)

These two kinds of studies intersect in the following way. Some neuroendocrinologists claim to be able to identify the biological determinants of human behaviors. Traditionally androcentric evolutionary stories tell us that the roots of some human behaviors—namely those exhibited in the division of labor by gender—are to be found in the history of human evolution. Some eminent scientists even propose that the origins of Western, middle-class social life today, in which men rule in the public realm and women perform domestic labor, are to be found in the bonding of "man-the-hunter" with other men to go off and kill large animals while the women supposedly stayed at home in the cave to nurture children.[14]

> If these broadly described behaviors or behavioral tendencies could be correlated with the more particularized behaviors and behavioral dispositions studied by neuroendocrinology, a picture of biologically determined human universals would emerge. Evolutionary studies would provide the universals—gender and sex roles that have remained fundamentally constant throughout the history of the species—while neuroendocrinology provided the biological determination—the dependence of these particular behaviors or behavioral dispositions on prenatal hormone distribution.[15]

Thus if the dominant hypotheses in either area are unsupportable, so is their conjunction: the biological determinist case requires plausible arguments both for the existence of sex-role behavioral universals across all cultures, *and* for the genetic origins of these behaviors in individuals. Neither set of hypotheses has been uncontroversial among biologists—

[14]See, e.g., Edward Wilson's *On Human Nature* (Cambridge, Mass.: Harvard University Press, 1978).
[15]Longino and Doell (1983, 223). Subsequent page references to this paper appear in the text.

feminist or not. But I shall focus on the evolutionary hypothesis, both for reasons of brevity and because the issues here are more quickly grasped by nonbiologists than are those in neuroendocrinology.

Helen Longino and Ruth Doell offer a useful schematization of the points at which evolutionary studies are vulnerable to charges of androcentric bias. I shall supplement their account with arguments made by other biologists. Following one schematized "history" of an inquiry in this way will provide not only examples of prevalent kinds of masculine bias but also a more detailed look at the inquiry process and the variety of ways in which cultural bias can influence the eventual results of research. Longino and Doell show that masculine bias can enter both evolutionary and endocrinological research at a number of different points: "what questions are asked; what kinds of data are available, relevant, and appealed to as evidence for different types of questions; what hypotheses are offered as answers to those questions; what the distance between evidence and hypothesis is in each category; and finally, how these distances are traversed" (p. 210).

For our purposes, however, Longino and Doell's analysis is interesting for reasons beyond their documentation of the entry points of bias. While they think feminist observers of science need not choose between criticizing bad science and science-as-usual, in fact they conceptualize the androcentrisms they report as primarily issues of bad science. They appear to believe that the methodological norms of biology are not problematic, that biology can be reformed to eliminate masculine bias. They proffer their analysis as a rebuttal to feminists who argue that there is something androcentric about scientific method itself.

In the evolutionary studies, the questions asked are about anatomical and behavioral evolution, and about the relation between the two: which anatomical developments affected which behavioral developments, and vice versa? Except, perhaps, insofar as they focus on the role of biologically determined sex differences in human evolution, such questions do not seem particularly androcentric—but that exception is a big one. Longino and Doell note it but do not think it particularly problematic: "Some feminist critics [such as Ruth Hubbard] have suggested that the entire category 'sex differences' is a fabrication supported by sexism and by analytic tendencies in science that emphasize distinctions over similarities. More modestly, it can be argued that the concept 'tomboy' [which appears in the neuroendocrinological studies] identifies but mystifies a slight difference in behavior among young women. . . . An alternative perspective might invent a name for young women who are not tomboys and seek the determinants of their peculiar behavior" (p. 226). As for the evolutionary studies, they could similarly argue that the nineteenth century language of courtship so common in descriptions of the sexual activities of apes and other animals as well as humans could be replaced by value-neutral language. But is the issue of the definition of what is found problematic and therefore in need of scientific explanation (sex differences vs., for instance, both sex similarities and species differences) resolved by substituting purportedly purely descriptive for obviously androcentric language? Couldn't biology use totally value-neutral language (if there is any) and yet still be androcentric in its selection and definition of research problems?

The data available to answer the anatomical (and physiological) questions come primarily from fossils, and there are relatively few available of the earliest hominids. But Longino and Doell point out that modern methods of dating such remains permit relatively reliable assignment of fossils to an evolutionary sequence. Also relatively reliable is the data base for conclusions about individual or noninteractive physical behaviors such as diet and locomotion.

Most controversy centers on "data relevant to the evolution of social, interactive behavior in its relation to the development of human anatomy" (p. 212). The data considered relevant here is from three sources: fossils—including the "estimated size and quantity of remains at hominid sites"; modern-day human hunter-gatherer societies; and existing primate societies. "Since there is considerable variation among human as well as nonhuman primate groups, the relevance of the observed behavior of any one of these societies to the reconstruction of the behavior of early hominids is constantly in question. . . . The behavior of contemporary apes, which represent an evolved rather than an original species, is, in any case, a questionable model for the behavior of our hominid ancestors" (p. 212). However, the fact that apes and modern hunter-gatherers are evolved species and different from hominids is not the only problem with using observations of ape behavior as evidence for generalizations about early or modern human cultures. Longino and Doell do not note that most available observations of apes, including very recent studies, have been collected by observers unaware of the need to avoid androcentrism. Consequently, these studies show a high tendency to project onto ape "nature" and social relations both racist and sexist projects of the observers' own societies.[16] Furthermore, humans are not the only species that can learn from experience and creatively adapt to changes in the environment. Selective collection, interpretation, and use of data about ape societies create the false image that ape social life is itself entirely biologically determined, thereby begging the question at issue.[17]

Anthropologists are similarly skeptical about the assumption that the social patterns of contemporary hunter-gatherer societies are the same as those of our ancestors at the dawn of human history. They show how even the earliest observations by Westerners, who presumed they had found humans untouched by Western development, were in fact observations of groups who had already been forced to adapt to the cultural patterns of the West. Eleanor Leacock, for example, argues that the masculine dominance that eighteenth-century Westerners described among the hunter-gatherer societies of Canada was *entirely* an artifact of the combination of two factors: androcentric expectations on the part of the observers (which affected not only the Westerners' selective collection and interpretation of observations but also the actual behavior that the hunter-gatherers chose to exhibit to these Westerners), and the adaptations these societies had already made as a result of the changes in their economic activities caused by the presence of Westerners in their vicinity.[18] She argues that assumptions of universal masculine dominance are faulty—that many cultures were gender-egalitarian prior to influence by the West. Not all anthropologists agree with those who make this argument, but the point stands as a corrective to the attribution of unevolved primitiveness to recently observed hunter-gatherer cultures.

Androcentric assumptions, then, appear far more common in the collection, interpretation, and use of data about the dawn of human history than Longino and Doell's account indicates. How much androcentrism can be eliminated by the creation of alternative accounts and by stricter adherence to the existing methodological standards of biological inquiry? While men as well as women have contributed to the criticisms of androcentric biology, what is the significance of the fact that the alternative accounts have been produced by *women* inquirers, and in the midst of the second wave of the women's movement?

[16] A particularly illuminating analysis of this problem is provided by Haraway (1978).
[17] See Leibowitz (1978).
[18] Leacock (1982).

Longino and Doell point out that feminists have developed a "more comprehensive and coherent" theory than the current "man-the-hunter" hypothesis (p. 216). "Man-the-hunter" has been claimed responsible for the development of tools as aids in hunting. This (presumably only masculine) tool use itself favored the development of bipedalism and upright posture and, consequently, more effective hunting strategies featuring greater cooperation through division of labor among the hunters. It also made possible changes in dentition, once men could display aggression "by brandishing and throwing objects rather than baring or using the canines," and these changes made possible more energy-effective diets.[19] In the hands of some defenders of this theory, men's hunting behavior is claimed to be the evolutionary origin of "male bonding" in contemporary society, and thus there are supposed to be good evolutionary reasons why men seek to exclude women from their economic activities—such as, presumably, science. Such a hypothesis presents men as the sole creators of the shift from prehuman to human cultures. Furthermore, the vast cultural distance between early human cultures and industrial capitalism is explained as entirely due to the continued elaboration of men's biological "imperatives" to create culture. The activities of women in contemporary societies (barring the activities of "unnatural women" such as feminists, of course) are presented as fundamentally the same as the activities of females in prehuman groups. As one biologist has suggested, this kind of account—of which Darwin also was guilty—creates the impression that were it not for the fortunate fact that daughters as well as sons inherit their father's genes, the mates of contemporary men would have to be female apes. As this biologist entitles an essay, "Have Only Men Evolved?"[20]

Longino and Doell discuss the alternative "woman-the-gatherer" hypothesis developed by some anthropologists.[21] Where man-the-hunter invented primarily stone tools, it is likely that women earlier invented tools made of organic materials such as sticks and reeds. This is hypothesized to be "a response to the greater nutritional stress experi-

---

[19]This argument was originally made by Sherwood Washburn and C. S. Lancaster, "The Evolution of Hunting," in Richard Lee and Irven DeVore, eds., *Man the Hunter* (Chicago: Aldine, 1968).

[20]Hubbard (1979).

[21]See Frances Dahlberg, ed., *Woman the Gatherer* (New Haven, Conn.: Yale University Press, 1981); Tanner (1981); Tanner and Zihlman (1976); Zihlman (1978). See Caulfield (1985) and Zihlman (1985) for evaluations of the implications and effects of feminist rethinkings of human evolution that appeared after the Longino and Doell paper but do not challenge its arguments.

97

enced by females during pregnancy, and later in the course of feeding their young through lactation and with foods gathered from the surrounding savannah" (p. 213). Other anthropologists argue that bipedalism caused what is known as the "obstetrical dilemma" in our evolutionary history: bipedalism narrowed the birth canal, while tool use created selective pressures for larger brain size and thus for larger craniums. The resolution to this dilemma was to have human infants born at a less mature stage than is characteristic of prehumans. Less mature infants require greater and longer adult—not necessarily women's—care (though this increased period of close association with adults also makes possible more extensive socialization of human neonates than is possible with the more mature neonates of prehumans), with probable increased nutritional stress in women as a result.

This gynecentric story of the origins of human culture portrays "females as innovators who contributed more than males to the development of such allegedly human characteristics as greater intelligence and flexibility. Women are said to have invented the use of tools to defend against predators while gathering and to have fashioned objects to serve in digging, carrying, and food preparation" (p. 213).

Which of these stories should we find plausible? Longino and Doell point out that the "distance between evidence and hypotheses" is less, but only slightly less, for the woman-the-gatherer than for the man-the-hunter hypothesis; that is, the former is slightly better supported by its evidence than the latter by its (different) evidence. Generalizations about the uses and users of tools are the means by which the path from evidence to hypothesis is traversed in each case, and analogies with contemporary populations of hunters and gatherers are used to support the generalizations. But as Longino and Doell note, "the behavior and social organization of these peoples is so various that, depending on the society one chooses, very different pictures of Australopithecus and Homo erectus emerge" (p. 215). The only tools that have been recovered are of stone, for of course the organic materials from which many of women's tools would have been fashioned are not to be found. But women may have used stones as well,

> to kill animals, scrape pelts, section corpses, dig up roots, break open seed pods, or hammer and soften tough roots and leaves to prepare them for consumption. . . . If female gathering behavior is taken to be the crucial behavioral adaptation, the stones are evidence that women began to develop stone tools in addition to the organic tools already in use for gath-

98

ering and preparing edible vegetation. If male hunting behavior is taken
to be the crucial adaptation, then the stones are evidence of the male
invention of tools for use in the hunting and preparation of animals. . . .
It is . . . a matter of choosing a male-centered or female-centered framework of interpretation and assigning evidential relevance to data on the
basis of that framework's assumptions. [p. 215]

There is no possibility for the chipped stones to become more convincing evidence for one or the other hypothesis, no possibility of
gathering additional evidence which could tip the evidential scales even
slightly one way or the other. What could possibly constitute such
evidence?[22] Nevertheless, even if we cannot now or perhaps ever make
a convincing case for one of these stories to the exclusion of the other,
Longino and Doell point out that the feminist criticisms have served
a useful function. They have revealed a number of points at which
androcentric bias has shaped the man-the-hunter theory. "Androcentric bias is expressed directly in the framework within which data are
interpreted: chipped stones are taken as unequivocal evidence of male
hunting only in a framework that sees male behavior as central not
only to the evolution of the species but to the survival of any group
of its members" (p. 224). The creation of an alternative framework
"may not provide the final word in evolutionary theorizing, but it does
reveal the epistemologically arbitrary nature of those androcentric assumptions and point the way to less restrictive understandings of human possibilities" (p. 225).

Longino and Doell also note that "the assumption of cross-species
uniformity and the adequacy of animal modeling is highly questionable
in its application to behavior" (p. 226). But they do not point out that
the particular forms of these assumptions in biological determinist
accounts appear to be not just generally questionable but androcentric.
As one biologist points out, it is more plausible to assume that human
men and women are far more similar to each other than any humans
are to members of other species. And what is uniform within the human
species is its immense plasticity, creativity, and conscious adaptability.

Even if some sex-associated behaviors *were* found to be universal among
all nonhuman primates or indeed among all mammalian species, generalizations to human behavior and social relationships would have to ignore

[22]See Harding (1976) for discussion of the different issue of whether there can in
principle be "crucial experiments."

five million years of exuberant evolutionary development of the human brain, which has resulted in a cerebral cortex quantitatively and qualitatively different from that of other primates. It is a cortex that provides for conceptualization, abstraction, symbolization, verbal communication, planning, learning, memory and association of experiences and ideas, a cortex that permits an infinitely rich behavioral plasticity and frees us, if we choose, from stereotyped behavior patterns. . . . Not only is there no universal behavioral trait or repertoire among our closest relatives, the nonhuman primates, to study as a "primitive" prototype or precursor model for human "nature," there is no *human nature*, no universal human behavioral trait or repertoire that can be defined, *except* for our tremendous capacity for learning and for behavioral flexibility.[23]

The point here is that if we ask which gendered humans have historically been concerned—indeed, obsessed—to distinguish themselves from members of the *other* gender, the answer is "men." Similarly, it is men who have been preoccupied with finding the continuities between men and males in other species, and between women and females in other species. Thus it is reasonable to believe that the selective focus on purported sexual samenesses across species and sexual differences within species is not only questionable but also a distinct consequence of androcentrism. It would certainly be unreasonable to regard it as an example of the pure (i.e., ungendered) intellect pursuing problems that nature creates for inquiry. Only masculine investment in the evolved distinctiveness of men's achievements and the unevolved naturalness of women's activities appears able to account for this excessive focus on samenesses between the species and differences between the sex/genders.

Thus Longino and Doell show how androcentrism has entered evolutionary theory in its selection of what counts as a scientific problem, its concepts and theories, its methods of gathering data and selecting what data to use as evidence, and its interpretations of results. As I noted earlier, if the evolutionary hypotheses are not plausible, than neither are the biological determinist claims that depend upon the conjunction of these hypotheses with those of neuroendocrinology.

But should these biases be regarded merely as an example of "bad science"? May they not, instead, be assessed as fundamentally characteristic of modern Western science?

Before I pursue this issue, it is worthwhile to note a kind of general

[23]Bleier (1979, 58–59).

conceptual confusion endemic to attempts to trace human behaviors to innate or genetic inheritances, as does the neuroendocrinological research. Critics of this half of the biological determinist argument point out that genetic inheritances constitute arrays of possibilities, and which of these possibilities will be expressed in behaviors or behavioral dispositions depends upon the environment within which the genes are located. "Behavior results from the joint operation of genes and the environment, and these factors interact in complex and nonlinear ways that are different and unpredictable for different traits."[24] Thus it is meaningless to try to partition genetic and environmental components in behavior and discuss them separately, as the biological determinists do.

> In the absence of knowledge about these interactions for each specific trait one wishes to consider—and at present we do not have it about a single observed human behavior—the only meaningful question that can be asked is how much of the observed *variation* in behavior among individuals is caused by genetic factors and how much by environment. ... This more limited question tells one *nothing* about how to partition genetic and environmental effects for the behavior itself; nor does it tell us anything about the proportion of genetic and environmental contributions to the variance of any other trait.[25]

Furthermore, even if we could partition the variation for any particular trait, this feat would not enable us to predict that the same partition would appear in a different environment; the relative contributions of genetic and environmental factors are likely to change when the environment changes. "Thus, in comparing two groups that differ genetically, it is impossible to distinguish the genetic and environmental origins of *any* behavioral differences between them as long as their environments differ in *any* way."[26] This leaves us with the fact that gender behaviors do differ in our culture as in others, and that in principle we cannot separate the genetic from the environmental causes of them. What we can do is try to show interactions between genetic inheritance and environmental conditions for historically specific behaviors. But that is quite different from the project of the biological determinists. As we shall later see, defensible boundaries between the

[24]Lowe and Hubbard (1983, 95).
[25]Lowe and Hubbard (1983, 95–96).
[26]Lowe and Hubbard (1983, 95).

concepts of human and nonhuman have irretrievably broken down for our culture. However, biological determinism is not the only available response to this emerging recognition.[27]

IS THE PROBLEM "BAD SCIENCE" OR "SCIENCE AS USUAL"?

Longino and Doell's essay is in part an attempt to resolve the paradox in which many feminist critics appear to ground their claims. More often than in the social sciences, feminist critics of biology simultaneously challenge as androcentric the entire scientific methodological ethos of objectivity, value-neutrality, dispassionate inquiry, and the like, and yet also claim to provide objective, value-neutral, dispassionate facts about nature and social life. On the one hand, feminists have used the Kuhnian strategy of arguing that observations are theory-laden, theories are paradigm-laden, and paradigms are culture-laden: hence there are and can be no such things as value-neutral, objective facts.[28] On the other hand, these very same critics present alternative descriptions and explanations of nature and social life as factual or true—not merely as differently culture-laden. (Of course, this paradox can be raised against Kuhn's own analysis.)

But feminists do not propose that androcentric and feminist accounts are explanatorily equal—that it would be equally reasonable to accept either—any more than Kuhn proposes that his account is only as plausible as the accounts he criticizes. We cannot accept as the last word in evolutionary theory *both* the "man the hunter" and the "woman the gatherer" hypotheses, for the two conflict. The feminist theorists think that the "man the hunter" hypothesis has masculine bias but that their account is not equally gender-biased; it is simply more plausible because it transcends the masculine gender bias of the traditional account. And they think this in spite of the fact that they were admittedly motivated to provide this account at least in part because they thought it was morally and politically wrong to devalue women's activities as the dominant account did. However, they think that every reasonable scientist should regard their account as more plausible on the evidential grounds—not because it originated in feminist moral and political concerns or because it privileges women's activities in the evolution of culture.

[27]Haraway (1985).
[28]Kuhn (1970).

102

Donna Haraway raises the issue of feminist vacillation between calling androcentric science "bad science" and calling it "science-as-usual"
without proposing a solution to the paradox.[29] But Longino and Doell
think that we are not forced to make this choice:

> If sexist science is bad science and reaches the conclusions it does because
> it uses poor methodology, this implies there is a good or better meth-
> odology that will steer us away from biased conclusions. On the other
> hand, if sexist science is science as usual, then the best methodology in
> the world will not prevent us from attaining those conclusions unless we
> change paradigms. . . . Feminists do not have to choose between correct-
> ing bad science or rejecting the entire scientific enterprise. The structure
> of scientific knowledge and the operation of bias are much more complex
> than either of these responses suggests. [pp. 207–8]

Longino and Doell's underlying insight in this passage is certainly
right: feminists cannot afford to leave correctable examples of masculine bias uncriticized, nor should we want to reject science entirely. But their otherwise useful account would lead us to believe
that androcentrism in science is entirely a consequence of ignorance
and faulty reasoning, that more balanced understanding of human
evolution will result if biologists—both men and women—simply look
at different evidence, construct different arguments, and use different language. While these activities certainly cannot do any harm, I
suggest that they will not result in the elimination of masculine bias.

The problem is that the Longino and Doell analysis misconceives
what biology is and what gender is. The offending biology is thought
of as a set of sentences and methodological procedures; if we substitute different sentences and methodological procedures for the androcentric ones, we will have substituted good science for bad science.
However, is it not more plausible to argue that a more robust conception of evolutionary biology would understand it as part of a
seamless weaving of our culture's dominant social projects? Attempts to add unbiased accounts of women's activities and of social
relations between the sexes confront not just inadvertent gaps and
distortions in the text of science but, more important, the seamless-
ness of science's participation in projects supporting masculine domination. The social projects of cultures in which scientific inquiry
occurs, as well as the ignorance and false beliefs of individual inquir-

[29]Haraway (1981). Haraway (1985) begins to move beyond this impasse.

ers, appear to be responsible for the selection of scientific problems, for the kinds of hypotheses proposed, for the determination of what is to count as evidence, and for how that evidence is taken to support or disconfirm hypotheses. As I emphasized earlier, gender is asymmetrically organized: part of what it means to become gendered as masculine is to become that kind of social person who is valued more highly than women. And we saw that this masculine identity is excessively fragile. Where women cannot seem to escape being perceived as feminine, men seem to fear that they will no longer be men unless they constantly prove their masculinity. Thus it is the masculinity-affirming division of labor by gender, the assignment of individual gender identities to human infants, and the asymmetric meanings of masculinity and femininity in gender symbolizing—in gender totemism—that create the androcentric biases in biology. We should not expect the feminist arguments to change the way biological theory is made and research is done—or even to be assessed as plausible by the majority of masculine-gendered biologists—until all three of these forms of gender begin to be eliminated. The demonstration of evils is rarely sufficient to eliminate them, as Margaret Rossiter pointed out.[30]

Both Longino and Doell's kind of analysis and that of the feminist evolutionary theorists whose work they examine make important—indeed, necessary—contributions to our attempts to degender science. But their potency for this task depends on our ability to grasp the reasons and causes of our having to undertake these kinds of projects at all. Is it possible to create a theory of how humans evolved from other species that does not project onto nature distinctive human projects? Perhaps it is not just the interest in "sex differences" that reflects distinctively masculine cultural interests but the very interest in human evolution. Still more generally, is not all biology, as the locus of the intersection of nature and culture, doomed to value-bias? If this line of thinking is plausible, we appear to be faced with the option either of accepting a fundamental cultural relativism in our biological explanations (in whichever explanations appear most plausible to us at any given moment in our history) or of conceptualizing objectivity in biological research in a way very different from its conceptualization in physics. Alternatively, as indicated earlier, we may need to rethink the causes and reasons for whatever objectivity physics has attained.

[30]Rossiter (1982b).

104

Longino and Doell are right: our choices need not be between bad science and rejecting all science. But in order to understand our more fruitful choices, we need to work through the feminist epistemological issues.

IMPLICATIONS

The kinds of bias discussed above appear, at first glance, to be found only in the social sciences and biology—not in chemistry, astronomy, or physics. Hence these problems appear peculiar to studies of social phenomena—including those of socially constructed and socially meaningful "bodies"—and may not appear to challenge the epistemology or politics of the scientific enterprise more generally. However, the apparent imperviousness of the exclusively physical sciences to charges of masculine bias does not weaken the importance for my project of the issues these studies raise.

In the first place, there *is* in fact masculine bias in fields of inquiry that have from their beginnings tried to achieve the kind of objectivity thought characteristic of physics. Such bias does not only appear in concrete issues of the limited access men can get to women's worlds, or of the invisibility of social analyses of women's worlds. It also appears in extremely abstract and therefore apparently innocent components of these sciences: in models of what constitutes the social order and distinctively cultural activities; in assumptions about the fit between social actors and the roles assigned to them; in the heretofore unnoticed and suspicious fit between the conceptual categories of theories and those of masculine informants; in the equally suspicious fit between the categories of social science and those of the administrators and managers of industrial capitalism; and perhaps even in assumptions about the relative importance of sex differences within species and sameness across species.

Moreover, it appears to be exactly within some attempts by social science to mimic the purportedly objectivity-increasing aspects of the physical sciences that feminists claim to find distinctively masculine bias. The social science criticisms have already led us to suspect that the focus on quantitative measures, variable analysis, impersonal and excessively abstract conceptual schemes is both a distinctively masculine tendency and one that serves to hide its own gendered character. Does the methodological preference in the social sciences for delineating hierarchical structures of the simplest kinds of differences rather than the reciprocal, interactive relations of complexes express a distorting masculine bias that appears in the natural sciences as well?

In the second place, biology and the social sciences have been key culprits in promulgating what we can now see as false and socially regressive understandings of women's and men's natures and "proper" activities in social life. Is it an accident that many of these biological and social theories were created in nineteenth-century Europe and America during a period of vast change in women's and men's traditional division of labor, during shifts in the meanings and referents of heterosexuality, and during the beginnings of agitation for equal education, employment opportunity, and political suffrage for women? The alternative feminist accounts have been constructed in the midst of similar changes and movements.

In the third place, the recognition that influential theories in the social and life sciences were invented in the midst of and as weapons in historical battles between the genders is interesting in its own right. But since the physical sciences, no less than biology and the social sciences, are historical creations—cultural artifacts—this recognition also raises suspicions about the imperviousness in principle of *any* scientific theory to influence by the gender order, as well as by race, class, and cultural hierarchies.

Fourth, one significant feature of the Millman and Kanter and the Longino and Doell studies is the inherent tension they exhibit between directions for the reform of the sciences we have and assumptions that run fundamentally counter to the epistemology of those sciences. Both studies state or imply that were feminists (not just "women") more evident in the design and execution of research, a more comprehensive picture of human activity would result. Women's access to different data from men's is itself an important source of improved science. But feminists (women and men) also tend to ask different questions, have different perceptions, and interpret data differently; and both studies suggest the positive effect of at least some kinds of politicized inquiry. The Millman and Kanter essay begins by noting:

> Movements of social liberation . . . make it possible for people to see the world in an enlarged perspective because they remove the covers and blinders that obscure knowledge and observation. In the last decade no social movement has had a more startling or consequential impact on the way people see and act in the world than the women's movement. . . .

106

We can see and plainly speak about things that have always been there, but that formerly were unacknowledged. Indeed, today it is impossible to escape noticing features of social life that were invisible only ten years ago. [p. vii]

Longino and Doell implicitly support this kind of analysis by their constant use of such terms as "feminist biologists," "feminist critics," "feminist alternative accounts." They specifically want to show that we are not faced with a choice between "bad science" and "science as usual." But in their assumption that the ethics and politics of the women's movement ("feminism") are at least in part responsible for the more comprehensive and coherent "woman-the-gatherer" theory, they support a logic of scientific inquiry diametrically opposed to that of the traditional accounts. As I noted earlier, the contemporary remnants of traditional philosophy of science distinguish between contexts of discovery and contexts of justification; the growth of knowledge is supposed to be advanced exclusively by rigorous justification procedures. But the studies we have been reviewing suggest that political motivations for inquiry, which are responsible for getting hypotheses up to the starting point for workover by the rigorous procedures of justification and which play a significant role in the selection of what constitutes rigorous procedure, may have a greater influence on what gets counted as justified belief than any purportedly value-free specification of method. An important origin of androcentric bias in social science and biology occurs in the context of discovery—in the selection and definition of problems for inquiry. What *is* the desirable relationship between science and politics if it is not the complete separation claimed by contemporary science enthusiasts? Is the genealogy of beliefs in both practice and principle an important factor in the legitimacy of their justifications? That is, do scientific beliefs achieve legitimation in part because of their social origins? Shouldn't the social origins of beliefs then be *one* factor to be considered in their justification? Isn't it reasonable to suppose that claims originating in racist and sexist projects may well be less worthy of scientific attention—less likely to "reveal reality"—than those originating in antiracist and antisexist projects? The familiar epistemologies of science explicitly reject this kind of supposition.

Finally, since natural science is itself a social enterprise, an adequate understanding of the history of science and its gender politics requires an adequate philosophy of social science—which is not forthcoming in the tendency to see the issues as how to improve and reform "bad science."

We have already seen that even the "least-threatening" feminist challenges to science, the affirmative action issues, point to the possibility that real achievement of equal opportunity for women requires a radical reduction in gender stereotyping, in the division of labor by gender, and in the defensive fragility of masculine identity—perhaps the complete elimination of gender and therefore of gender stratification in societies that produce science. Now we are led to suspect that the next least threatening feminist challenges, the elimination of masculine bias in social science and biological theory and research, requires a fundamental transformation of concepts, methods, and interpretations in these areas, and a critical examination of the logic of scientific inquiry—rather more than mere reforms.

We can see that the Woman Question criticisms, when thought to be asking merely for reforms in scientific practices, are conceptualizing women as a special-interest group with overlooked needs and interests—like children, or the differently abled, or farmers—which a democratic society has the moral (but not epistemological) obligation to accommodate. Perhaps people present and perceive the feminist criticisms we have examined in this way because interest-group politics is a recognized and legitimate form of political negotiation in our society. Interest-group politics assumes that individuals who happen to have distinctive interests have a moral and political right to be recognized in a pluralistic society, but that this right is limited to those who do not propose the overthrow of the ideas and institutions of democratic, pluralistic, interest-group politics. And since science appears explicitly to embrace such ideals, the Woman Question criticisms are not perceived as challenging the political model of science.

However, this kind of thinking makes puzzling the historical and continuing resistance of the scientific establishment to these feminist criticism. Is such resistance due simply to expectable reluctance to give up familiar patterns of behavior and the concepts and theories that men have devoted their careers to defending? Or is more at stake than career patterns?

This kind of thinking also distorts the feminist claims. Feminists are not arguing that anti-sexist theory, research, and politics have an *equal right* to be recognized as legitimate or desirable *alongside* sexist theory, research, and politics. They are not arguing that women should be granted the dubious gift of being permitted to work alongside colleagues and within institutional norms and practices that are obviously sexist or that women should have to "become men" (that is, take on masculine personalities and life patterns) in order to practice science. They are not arguing that antisexist and sexist problematics, concepts, theories, methods, and interpretations should be regarded as scientifically *equal*. They are arguing outside this kind of pluralistic politics, for reasons that should be obvious. Sexist science is morally and politically wrong because it supports those desires and interests of men that are satisfied only at the expense of women as a group. Individuals do not happen to be women or men by biological fiat; they are constituted as gendered by identifiable social processes. And this pluralism is scientifically wrong because it hides real regularities and underlying causal tendencies in social relations and relations between humans and nature. Is science's interest-group politics an obstacle to adequate understanding of nature and social life?

These claims should not be taken to support the idea that every claim a woman makes or every claim made in the name of feminism is thereby automatically more legitimate, politically and scientifically, than the understandings otherwise produced. In fact it is very difficult in most specific cases to decide what claims are best supported by moral and political or by scientific reason and evidence. And the gender of the claimant is often irrelevant to the kind of reason and evidence the claim can in principle gather. After all, many men have made outstanding contributions to the feminist theory and politics of their day (think of Plato, Karl Marx, John Stuart Mill, and Frederick Douglass, as well as the many contemporary men who are feminist scholars), and at least some women have made notorious contributions to sexist theory and politics (think of Anita Bryant, Marabel Morgan, Phyllis Schlafly). And while I have been using the term "feminism" here as if it were a monolithic set of beliefs and practices, it is not; there are significant differences among feminists about what analyses and practices are desirable (differences which for the most part are an important resource for future theory and politics). But when we are satisfied that reason and evidence do provide support for a feminist claim, that claim is intended to replace—not to coexist on an equal footing with—androcentric claims.

If the feminist criticisms can no longer be seen simply as demands that the social sciences and biology adhere more rigorously to their own directives for objective, value-neutral inquiry—if those directives are themselves suspected to be an expression of androcentrism—feminist inquiry has apparently grounded its claims in a paradox. Clearly, more scientifically rigorous and objective inquiry has produced the evidence supporting specific charges of androcentrism—but that same inquiry suggests that this kind of rigor and objectivity is androcentric! It is this paradox that raises the Science Question in feminism.

5 NATURAL RESOURCES: GAINING MORAL APPROVAL FOR SCIENTIFIC GENDERS AND GENDERIZED SCIENCES

Feminist criticisms have problematized the fact that for more than three centuries science has both explicitly and implicitly appealed to gender politics as a moral and political resource for its own advancement. Equally problematic is the recognition that those interested in gender politics have in turn consistently appealed to science to naturalize the subjugation of women. This kind of mutual support can also be detected in the symbiotic relationship between racism, classism, and science.

This chapter looks at how one form of gender politics—gender symbolism—has provided resources for the moral and political advancement of scientific modes of knowledge-seeking, and at how science has in turn advanced modern forms of gender symbolism. As we go along, we shall remind ourselves that gender symbolism is always supported by either actual divisions of labor by gender or perceived threats to existing gender-divided activity, and that it also has a complex relationship with individual sex and gender identities and prescribed behaviors. That is, gender symbolism rarely reflects in an undistorted manner a culture's divisions of labor or its participants' sex and gender identities. Lest any reader entertain lurking suspicions that the gender symbolism at issue is in fact simply an empirically supported report of the way the world is, I shall also review recent literature that reveals the social construction not only of gender but of much of what is commonly meant by the term "sex differences."

SHOULD THE HISTORY AND PHILOSOPHY OF SCIENCE BE X-RATED?

This question[1] is only slightly antic once we look at the metaphors and models of gender politics with which scientists and philosophers of science have explained how we all should think about nature and inquiry. Examples of gender symbolization generally occur in the margins, in the asides, of texts—in those places where speakers reveal the assumptions they think they do not need to defend, beliefs they expect to share with their audiences. We will see assumptions that the audiences for these texts are men, that scientists and philosophers are men, and that the best scientific activity and philosophic thinking about science are to be modeled on men's most misogynous relationships to women—rape, torture, choosing "mistresses," thinking of mature women as good for nothing but mothering. Let us look first at some striking examples from the history of science, and then examine some comments by contemporary scientists and philosophers.

Historical Images.

Contemporary science presents its conceptions of nature and inquiry as truths discovered at the birth of modern science—as objective, universally valid reflections of *the* way nature is and *the* way to arrive at mirrorlike descriptions and explanations. But historians point out that conceptions of nature and inquiry have changed over time, and that they have been highly influenced by the political strategies used in historically identifiable battles between the genders. Gender politics has provided resources for the advancement of science, and science has provided resources for the advancement of masculine domination. I raised this issue earlier in asking whether it could possibly be reasonable to regard as a pure coincidence the development of sexology hot on the heels of the nineteenth-century women's movement.

We should note at the start that there are a number of problems with these historical studies. One origin of these problems is the mystifying philosophy of social science directing them, especially the misleading understandings of the complete "life history" of the role of metaphor in scientific explanation. Another origin is the inadequacy of histories which say little about social relations between the genders,

[1]My apologies to Stephen Brush, whose paper, "Should the History of Science Be X-Rated?" in *Science* 183(no.4130) (1974) did not deal with the gender behavior of scientists (or philosophers).

112

let alone about how changes in these relations were experienced, perceived, and responded to by the culture in general, including the scientific thinkers of the day. We can see that the five substantive problems with the conceptual schemes of the social sciences pointed out by feminist critics (see Chapter 4) infest the source materials available to historians today. In spite of such shortcomings, these studies greatly advance our understanding of science's place in its social worlds.

One phenomenon feminist historians have focused on is the rape and torture metaphors in the writings of Sir Francis Bacon and others (e.g., Machiavelli) enthusiastic about the new scientific method. Traditional historians and philosophers have said that these metaphors are irrelevant to the *real* meanings and referents of scientific concepts held by those who used them and by the public for whom they wrote. But when it comes to regarding nature as a machine, they have quite a different analysis: here, we are told, the metaphor provides the interpretations of Newton's mathematical laws: it directs inquirers to fruitful ways to apply his theory and suggests the appropriate methods of inquiry and the kind of metaphysics the new theory supports.[2] But if we are to believe that mechanistic metaphors were a fundamental component of the explanations the new science provided, why should we believe that the gender metaphors were not? A consistent analysis would lead to the conclusion that understanding nature as a woman indifferent to or even welcoming rape was equally fundamental to the interpretations of these new conceptions of nature and inquiry. Presumably these metaphors, too, had fruitful pragmatic, methodological, and metaphysical consequences for science. In that case, why is it not as illuminating and honest to refer to Newton's laws as "Newton's rape manual" as it is to call them "Newton's mechanics"?

We can now see that metaphors of gender politics were used to make morally and politically attractive the new conceptions of nature and inquiry required by experimental method and the emerging technologies of the period. The organicist conception of nature popular in the medieval period—nature as alive, as part of God's domain—was appropriate neither for the new experimental methods of science nor for the new technological applications of the results of inquiry. Carolyn Merchant identifies five changes in social thought and experience in Europe during the fifteenth to seventeenth centuries that contributed

[2]See, e.g., the philosophers and scientists criticized in Hesse (1966).

to the distinctive gender symbolism of the subsequent scientific world view.[3]

First of all, when Copernican theory replaced the earth-centered universe with a sun-centered universe, it also replaced a woman-centered universe with a man-centered one. For Renaissance and earlier thought within an organic conception of nature, the sun was associated with manliness and the earth with two opposing aspects of womanliness. Nature, and especially the earth, was identified on the one hand with a nurturing mother—"a kindly, beneficent female who provided for the needs of mankind in an ordered, planned universe"—and on the other with the "wild and uncontrollable [female] nature that could render violence, storms, droughts, and general chaos" (p. 2). In the new Copernican theory, the womanly earth, which had been God's special creation for man's nurturance, became just one tiny, externally moved planet circling in an insignificant orbit around the masculine sun.

Second, for the Platonic organicism, active power in the universe was associated with the alive, nurturing mother earth; for the Aristotelian organicism, activity was associated with masculinity and passivity with womanliness. Central to Aristotle's biological theory, this association was revived in sixteenth-century views of the cosmos, where "the marriage and impregnation of the 'material' female earth by the higher 'immaterial' celestial masculine heavens was a stock description of biological generation in nature." Copernicus himself draws on this metaphor: "Meanwhile, the earth conceives by the sun and becomes pregnant with annual offspring" (p. 7). Resistance to this shift in the social meaning of womanliness is evident in the sixteenth-century conflicts over whether it was morally proper to treat mother earth in the new ways called for by such commercial activities as mining. But as the experience of "violating the body" of earth became increasingly more common during the rise of modern science and its technologies, the moral sanctions against such activities provided by the older organic view slowly died away. Simultaneously, a criterion for distinguishing the animate from the inanimate was being created. (This distinction is a theoretical construct of modern science, not an observational given familiar to people before the emergence of science. And, as we shall see, it is one that increasingly ceases to reflect "common sense.") Thus

---

[3]Merchant (1980). Subsequent page references to this work (and the authors cited within it) appear in the text.

a "womanly" earth must be only passive, inert matter and indifferent to explorations and exploitations of her insides.

Third, the new universe that science disclosed was one in which change—associated with "corruption," decay, and disorder—occurred not just on earth, as the Ptolemaic "two-world view" held, but also throughout the heavens. For Renaissance and Elizabethan writers, these discoveries of change in the heavens suggested that nature's order might break down, leaving man's fate in chaos (p. 128). Thinkers of the period consistently perceived unruly, wild nature as rising up against man's attempts to control his fate. Machiavelli appealed to sexual metaphors in his proposition that the potential violence of fate could be mastered: "Fortune is a woman and it is necessary if you wish to master her to conquer her by force; and it can be seen that she lets herself be overcome by the bold rather than by those who proceed coldly, and therefore like a woman, she is always a friend to the young because they are less cautious, fiercer, and master her with greater audacity" (p. 130).

Fourth, man's fate seemed difficult to control because of disorder not only in the physical universe but also in social life. The breakdown of the ancient order of feudal society brought the experience of widespread social disorder during the period in which the scientific world view was developing. Particularly interesting is the possibility that women's increased visibility in public life during this period was perceived as threatening deep and widespread changes in social relations between the genders. Women were active in the Protestant reform movements of northern Europe, and Elizabeth I occupied England's throne for an unprecedentedly long reign. Prepared by the organic view's association of wild and violent nature with one aspect of the womanly, and by the absence of clear distinctions between the physical and the social, the Renaissance imagination required no great leap to associate all disorder, natural and social, with women. By the end of the fifteenth century, this association had been fully articulated in the witchcraft doctrines. To women was attributed a "method of revenge and control that could be used by persons both physically and socially powerless in a world believed by nearly everyone to be animate and organismic" (p. 140).

Fifth, the political and legal metaphors of scientific method originated at least in part in the witchcraft trials of Bacon's day. Bacon's mentor was James I of England, a strong supporter of antifeminist and antiwitchcraft legislation in both England and Scotland. An obsessive

115

focus in the interrogations of alleged witches was their sexual practices, the purpose of various tortures being to reveal whether they had "carnally known" the Devil. In a passage addressed to his monarch, Bacon uses bold sexual imagery to explain key features of the experimental method as the inquisition of nature: "For you have but to follow and as it were hound nature in her wanderings, and you will be able when you like to lead and drive her afterward to the same place again. . . . Neither ought a man to make scruple of entering and penetrating into those holes and corners, when the inquisition of truth is his whole object—as your majesty has shown in your own example" (p. 168). It might not be immediately obvious to the modern reader that this is Bacon's way of explaining the necessity of aggressive and controlled experiments in order to make the results of research replicable!

As I indicated earlier, this kind of analysis raises a number of problems and challenges, some of which we shall examine further in later chapters. There does, however, appear to be reason to be concerned about the intellectual, moral, and political structures of modern science when we think about how, from its very beginning, misogynous and defensive gender politics and the abstraction we think of as scientific method have provided resources for each other. The severe testing of hypotheses through controlled manipulations of nature, and the necessity of such controlled manipulations if experiments are to be repeatable, are here formulated by the father of scientific method in clearly sexist metaphors. Both nature and inquiry appear conceptualized in ways modeled on rape and torture—on men's most violent and misogynous relationships to women—and this modeling is advanced as a reason to value science. It is certainly difficult to imagine women as an enthusiastic audience for these interpretations of the new scientific method.

If appeal to gender politics provides resources for science, does appeal to science provide resources for gender politics? Do not metaphors illuminate in both directions? As nature came to seem more like a machine, did not machines come to seem more natural? As nature came to seem more like a woman whom it is appropriate to rape and torture than like a nurturing mother, did rape and torture come to seem a more natural relation of men to women? Could the uses of science to create ecological disaster, support militarism, turn human labor into physically and mentally mutilating work, develop ways of controlling "others"—the colonized, women, the poor—be just misuses of applied science? Or does this kind of conceptualization of the char-

116

acter and purposes of experimental method ensure that what is called bad science or misused science will be a distinctively masculinist science-as-usual? Institutions, like individuals, often act out the repressed and unresolved dilemmas of their infancies. To what extent is the insistence by science today on a value-neutral, dispassionate objectivity in the service of progressive social relations an attempt by a guilty conscience to resolve some of these early but still living dilemmas?

The history of biology and medicine reveal similarly striking uses of gender symbolism to reconceptualize nature—a project that naturalized gender politics as it genderized biology and medicine. L. J. Jordanova's study of eighteenth- and nineteenth-century biomedical science in France and Britain found that "sex roles were constituted in a scientific and medical language, and, conversely, the natural sciences and medicine were suffused with sexual imagery."[4] Science and medicine were fundamental to the Enlightenment writers' critical examination of social organization in three ways:

> First, natural philosophers and medical writers addressed themselves to phenomena in the natural world such as reproduction and generation, sexual behaviour, and sex-related diseases. Second, science and medicine held a privileged position because their methods appeared to be the only ones which would lead away from religious orthodoxy and towards a secular, empirically based knowledge of the natural and social worlds. Finally . . . science and medicine as activities were associated with sexual metaphors which were clearly expressed in designating nature as a woman to be unveiled, unclothed and penetrated by masculine science. [p. 45]

Consciously or unconsciously, Enlightenment thinkers refused to detach women's and men's social roles from the description and depiction of physiological differences. One striking and influential expression of this socialized biomedicine appears in the wax models of human figures used for making anatomical drawings and for educational display in popular museums.

> The female figures are recumbent, frequently adorned with pearl necklaces. They have long hair, and occasionally they have hair in the pubic area also. These "Venuses" as they were significantly called lie on velvet or silk cushions, in a passive, almost sexually inviting pose. Comparable male figures are usually upright, and often in a position of motion. The

---

[4]Jordanova (1980, 42). Subsequent page references to this essay appear in the text.

female models can be opened to display the removable viscera, and most often contain a foetus, while the male ones are made in a variety of forms to display the different physiological systems.... Not only is the literal naturalness of women portrayed, in their total nakedness and by the presence of a foetus, but their symbolic naturalness is implied in the whole conception of such figures. Female nature had been unclothed by male science, making her understandable under general scrutiny. [p. 54]

This image "was made explicit in the statue in the Paris medical faculty of a young woman, her breasts bare, her head slightly bowed beneath the veil she is taking off, which bears the inscription 'Nature unveils herself before Science' " (p. 54). Anatomically, males were depicted as representing active agents of our species, females as the objects of human (masculine) agency. Women's bodies were simultaneously presented as objects of scientific curiosity and as objects of (socially constructed) sexual desire.

Particularly interesting is the fact that women's social and occupational roles during the period were very diverse, not limited to those prescribed by the stereotypes. Everyone would have experienced this diversity in women's activities—including medical and scientific men of the period—so it cannot be that such gender symbolism was simply a passive reflection of an existing division of labor by gender in the social world around them. Instead, "the lack of fit between ideas and experience clearly points to the ideological function of the nature/culture dichotomy as applied to gender. This ideological message was increasingly conveyed in the language of medicine" (p. 42). Thus biomedical science intensified the cultural association of nature with passive, objectified femininity and of culture with active, objectifying masculinity—and was in return more intensely masculinized by this project.

Examination of more recent periods of escalating appeal to gender politics suggests that the intensified expressions of misogyny in the sciences of earlier periods were not representations of a free-floating overt misogyny that had the good fortune to encounter a resource in emerging scientific projects of the day; more likely, fundamental social changes between the genders were occurring or threatening to occur. Overt misogynous expression is best thought of as masculine protest literature; after all, one does not bother to state what is obvious or to agitate for something one already has. From this perspective, the relative lack of overt misogynous expression in other historical periods cannot be taken as a simple indicator of equality between the sexes (although a great deal more equality has existed at other times and places than in the last few centuries of Western life); rather, the lack of male protest often accompanies the relatively stable powerlessness of women and should therefore be taken as an indication of men's "distance from the problem."[5]

Contemporary Images.
The regendering of nature and inquiry was not a project only of the comfortably distant centuries. Prodigious energies have been put into projects of this sort right up through the present day.[6] Many commentators have suggested that notions familiar in popular and scholarly discussions of science are at least subliminally drawing on gender symbols. Common examples are such dichotomies as "hard" and "soft" data, the "rigor" of the natural science vs. the "softness" of social

[5]This kind of analysis can also be used to illuminate the reasons for different levels of overt sexism in the different strata of contemporary society. The model of the sexist projected by much social science research, as well as by such cultural figures as the male chauvinist or machismo latino, is a working-class person who overtly expresses his hostility to women and his ignorance about them; in comparison, the middle class, to which most social scientist themselves belong, appears relatively unprejudiced and tolerant. Yet it is not latinos and working-class men who design and direct the institutions that maintain the subjugation of women. Class stratification of overt sexism is better understood as a function of two other phenomena. In the first place, middle-class people are increasingly taught not to express sexism overtly. More important, men already established in the elite strata of the government and the professions are not personally threatened by affirmative action directives, and they can afford alimony and child support (even if they resent it). Working-class men and men in entry-level professional jobs feel the effect of attempts to gain equality for women far more than do the elite. Thus not only is "tolerance" taught to the middle class; it is a luxury they can afford. In making these points I am indebted to David Wellman's analysis of class variability in expressions of racism in "Prejudiced People Are Not the Only Racists in America" (1977, ch.1). Wellman's work is also valuable for its insistence that racism is fundamentally a structural feature of societies, which in turn produce racist "prejudices" as defensive attempts to "explain" the easily perceived gap between democratic ideology and the realities of racial stratification. This is the kind of analysis feminists should make of sexism. It should occasion feminist thinking about racism within feminism; moreover, it would predict an *increase* in sexist attitudes (even if not always overtly expressed) as the women's movement brings increased public awareness of the contradiction between gender stratification and our "democratic" ideals. The notion of a masculine "backlash," which is often invoked to account for the apparent recent escalation of pornography, rape, incest, wife-battering, and other overt expressions of hostility, is on the right track but not quite complex enough to capture the social dynamic that Wellman's account suggests. The feminist debate over pornography especially could benefit from this kind of analysis: pornography is a *solution* to some men's dilemmas, not a cause of them.

[6]Fee (1980); Hall (1973–74); Griffin (1978); Keller (1984); Bloch and Bloch (1980).

science, reason and intuition, mind and matter, nature and culture, and so forth, as well as familiar appeals to the "penetrating thrust of an argument," "seminal ideas," and the like. But let us take a look at some more extended and conscientious efforts at gender symbolism.

Consider the following conclusion to a recent Nobel Lecture where the laureate, a physicist, is summing up the history of his prizewinning work:

> That was the beginning, the idea seemed so obvious to me and so elegant that I fell deeply in love with it. And, like falling in love with a woman, it is only possible if you do not know much about her, so you cannot see her faults. The faults will become apparent later, but after the love is strong enough to hold you to her. So, I was held to this theory, in spite of all difficulties, by my youthful enthusiasm.... So what happened to the old theory that I fell in love with as a youth? Well, I would say it's become an old lady, who has very little that's attractive left in her, and the young today will not have their hearts pound when they look at her anymore. But, we can say the best we can for any old woman, that she has become a very good mother and has given birth to some very good children. And I thank the Swedish Academy of Science for complimenting one of them.[7]

And here is the closing passage of a widely cited paper by an eminent contemporary philosopher of science; the author, Paul Feyerabend, is explaining why his proposal for a rational reconstruction of the history of science is preferable to Karl Popper's: "Such a development, far from being undesirable, changes science from a stern and demanding mistress into an attractive and yielding courtesan who tries to anticipate every wish of her lover. Of course, it is up to us to choose either a dragon or a pussy cat for our company. I do not think I need to explain my own preferences."[8] The two passages present two cultural images of manliness: the good husband and father, and the sexually competitive, locker-room jock.

Even the position in the texts of these contemporary moral appeals to gender politics is illuminating. Each occurs as the final statement—as the summary thoughts the audience/readers are to take away with them. In case they hadn't noticed the reinforcement to masculinity of

[7] Richard Feynman, *The Feynman Lectures in Physics* (Reading, Mass.: Addison-Wesley, 1964), cited in Traweek (1986).

[8] Paul Feyerabend, "Consolations for the Specialist," in Lakatos and Musgrave (1970, 229).

the "purely cognitive" claims, each author drives his point home in his final message. The scientist and the philosopher are, indeed, men (in spite of their successes in cerebral careers? do men, too, fear certain kinds of success?); the audience likewise. Their partners—science and its theories—are exploitable women. A proposal should be appreciated *because* it replicates gender politics.

Evelyn Fox Keller points out that it is not just a few scientists and philosophers who project a defensive masculinity onto their activities. Even though the scientist is perceived as supermasculine, he is also thought to be less sexual than men in certain other occupations. A study of English schoolboys for instance reveals the following set of attitudes: "The arts are associated with sexual pleasure, the sciences with sexual restraint. The arts man is seen as having a good-looking, well-dressed wife with whom he enjoys a warm sexual relation; the scientist as having a wife who is dowdy and dull, and in whom he has no physical interest. Yet the scientist is seen as masculine, the arts specialist as slightly feminine."[9] Keller notes that the perception of science as "antithetical to Eros" is related to the perception of science as a supermasculine activity, and that both images can be found in early thinkers: " 'Let us establish a chaste and lawful marriage between Mind and Nature,' Bacon writes, thereby providing the prescription for the birth of new science. This prescription has endured to the present day—in it are to be found important clues for an understanding of the posture of the virgin groom, of his relation toward his bride, and of the ways in which he defines his mission."[10]

Keller argues that it is in the association of competence with mastery and power, of mastery and power with masculinity, and of this constellation with science that the intellectual structures, ethics, and politics of science take on their distinctive androcentrism. Such images simultaneously construct the institutionalized ethos of gendered sexuality and of science and, consequently, of the practices structured by these institutions. Science reaffirms its masculine-dominant practices and masculine dominance its purportedly objective scientific rationale through continual mutual support. Not only is this set of associations objectionable because it is sexist; it also makes bad science. It leads to false and oversimplified models of nature and inquiry that attribute power relations and hierarchical structure where none do or need exist.

[9]L. Hudson, *The Cult of the Fact* (New York: Harper & Row, 1972), p. 83, cited in Keller (1978, 189).
[10]Keller (1978, 190).

121

Keller sees alternative images and practices within the history of science that are respectful of nature's own complexity, not so closely tied to distinctively masculine identity projects, and more androgynous: "We need not rely on our imagination for a vision of what a different science—a science less restrained by the impulse to dominate—might be like. Rather, we need only look to the thematic pluralism in the history of our own science as it has evolved."[11] Keller points to many non-macho elements in the history of science. One of the themes of her intellectual biography of Barbara McClintock is the transcendence of gender in McClintock's scientific problematic, concepts and theory, and methods of research. McClintock's "feeling for the organism," her respect for the complexity of difference between individuals, her need to "listen to the material" all exemplify non-masculine tendencies that can also be detected elsewhere in the history of science. McClintock's work does not provide a feminist science, Keller argues, exactly because it transcends gender (though McClintock may have been more easily led to a deviant formulation of molecular biology, Keller speculates, because of her own status as a woman, as an outsider, a deviant, within science).[12]

But here Keller mistakenly indentifies feminism with the exaltation of feminine identity projects, rather than with exactly that transcendence of gender. While some feminists have engaged in a kind of "reverse discrimination" here, the majority have been critical of such tendencies.[13] Furthermore, Keller replicates traditional internalist history in stressing pluralism in the intellectual history of science while ignoring the social, political, psychological, and economic constraints that explain why some scientific ideas gain social legitimacy and others do not. There are social as well as intellectual reasons why "master molecule" theories gain ascendancy at one moment in history and interactive models at another. While these criticisms name real challenges for Keller's kind of account, they are certainly not peculiar to her approach to these issues. And it is difficult to imagine what could constitute evidence against her claim that notions of mastery and competence, masculinity, and science stand in mutually supportive relationships that are detrimental both to science and to women. (And, we might add, to men, who are asked to fulfill a demanding and distorting set of prescriptions for achieving maturity.)

[11]Keller (1982, 602).
[12]Keller (1983).
[13]Fee (1984).

122

Merchant, Jordanova, and Keller join a series of others who have focused on the conceptual dichotomizing central to scientific ideology and practice. Is this tendency itself a distinctively masculine one? Some critics argue that its roots are to be found in Judaism and Christianity, capitalism and colonialism, the European culture of the fifteenth through seventeenth centuries and its liberal political theory. Chapter 7 examines problems with the way feminists have conceptualized these dichotomies, but let us look now at what they have to say about them.

Like Merchant, Jordanova, and Keller, Elizabeth Fee argues that such dichotomies are distinctively masculine. She points out that while they can be detected in the ideology of gender in modern liberal philosophy, they must have far older roots, since they are evident in the entire history of Western philosophy.

The construction of our political philosophy and views of human nature seem to depend on a series of sexual dichotomies involved in the construction of gender differences. We thus construct rationality in opposition to emotionality, objectivity in opposition to subjectivity, culture in opposition to nature, the public realm in opposition to the private realm. Whether we read Kant, Rousseau, Hegel, or Darwin, we find that female and male are contrasted in terms of opposing characters: women love beauty, men truth; women are passive, men active; women are emotional, men rational; women are selfless, men selfish—and so on and on through the history of western philosophy. Man is seen as the maker of history, but woman provides his connection with nature; she is the mediating force between man and nature, a reminder of his childhood, a reminder of the body, and a reminder of sexuality, passion, and human connectedness. She is the repository of emotional life and of all the nonrational elements of human experience. She is at times saintly and at times evil, but always she seems necessary as the counterpoint to man's self-definition as a being of pure rationality.[14]

Fee argues that the insistence on these masculinist dichotomies is crucial in four ways to the maintenance of the belief that science is objective. First, issues about the production of knowledge must be kept distinct from those about the social uses of knowledge lest scientists be forced to take responsibility for goals beyond the pursuit of

[14]Fee (1981, 11–12). See also Carol Gould's "The Woman Question: Philosophy of Liberation and the Liberation of Philosophy," Caroline Whitbeck's "Theories of Sex Differences," and Anne Dickason's "Anatomy and Destiny: The Role of Biology in Plato's Views of Women," all in Gould and Wartofsky (1976); Griffin (1978).

123

knowledge for its own sake, and lest the public be encouraged to seek more power over the choice of what research is to be funded and who is to perform it.

Second, thinking and feeling must be kept separate lest scientific rationality be forced to respond to how people feel about the probable social consequences of their own or others' successful research on weapons, biomedical projects, and social control. "The roles of scientist and of citizen are distinct, and the scientist need feel socially responsible or emotionally involved only in her or his role as private citizen."[15] Historians point out that this shift of the domain of morality to private life is a modern invention. For Aristotle and the Greeks, it was in *public* life that the highest exercise of morality could be achieved. Science has gained status through its paradigmatic role as the institution where this separation of rationality from social commitment is most effectively policed, while the spread of scientific rationality to all the institutions of modern life leaves science in the powerful position of enforcing this separation in other areas of social life.

Third, the scientific subject, the scientist, must be kept separate from the scientific object—what he or she studies. As Merchant and Jordanova pointed out, the knowing mind is active but the object of knowlege is passive. It is the scientific subject's voice that speaks with general and abstract authority; the objects of inquiry "speak" only in response to what scientists ask them, and they speak in the particular voice of their historically specific conditions and locations.

Fourth, science must be presented as separate from society precisely to obscure its intimate relationships to political power.

> We are told that the production of scientific knowledge must be independent of politically motivated interference or direction. Yet we see scientists constantly testifying before congressional committees, we find scientists in law courts, and involved in disputes at every level of public policy. It is obvious that the experts take sides. It is also obvious that these "experts" are very often funded by corporate interests, and that there are few penalties for those who find that their research supports the positions of these powerful lobbies.[16]

Ruth Hubbard has also argued that this kind of dichotomizing reveals the intellectual, moral, and political projects of the science we

[15]Fee (1981, 18).
[16]Fee (1981, 19–20).

124

have to be sexist, classist, and racist.[17] Hubbard stresses science as a social construction, a historical enterprise that tells stories about us and the world around us. As a biologist, she has focused on the historical stories a classist, racist, and masculine-dominant social order has chosen to tell about sex differences. In analyses whose topics range from the writings of Darwin and other eminent men of science through contemporary biology texts, she shows the sexist, classist, and racist political projects supported by the maintenance of these kinds of dichotomies in sex-difference research. She argues that the very focus on sex differences in the face of the incredible similarities between the sexes may itself be a reflection of distinctively masculine projects.

Mind vs. nature and the body, reason vs. emotion and social commitment, subject vs. object and objectivity vs. subjectivity, the abstract and general vs. the concrete and particular—in each case we are told that the former must dominate the latter lest human life be overwhelmed by irrational and alien forces, forces symbolized in science as the feminine. All these dichotomies play important roles in the intellectual structures of science, and all appear to be associated both historically and in contemporary psyches with distinctively masculine sexual and gender identity projects. In turn, gender and human sexuality have been shaped by the projects of this kind of science.

Our title question for this section should now appear less surprising. Should this history and philosophy of science be X-rated? The sexist meanings of scientific activity were evidently crucial resources through which modern science gained cultural acceptance; they remain the resources that contemporary scientists and philosophers use to justify and explain their activities. They also are used to attract young people (young men, presumably) into science and the philosophy of science. How can this be "socially progressive"? As historian Joan Kelly-Gadol asks, once we understand women's situation to be as fully social as men's, must we not reevaluate purportedly progressive movements in Western history for their impact on women as well as on men—for their impact on "her" humanity as well as "his"?[18] Why should we regard the emergence of modern science as a great advance for humanity when it was achieved only at the cost of a deterioration in social status for half of humanity? Why should we regard the miso-

---

[17]Hubbard, Henifin, and Fried (1982); Hubbard (1979); Lowe and Hubbard (1983). Hubbard and Lowe are also listed as the editors for Tobach and Rosoff (1979), vol. 2 in the *Genes and Gender* series.

[18]Kelly-Gadol (1976).

gynous arguments of contemporary Nobel laureates and eminent philosophers of science as irrelevant to the meanings science has for scientists and the general public—especially when we are asked to understand other kinds of metaphors in science as intrinsic to the "growth of knowledge"? It seems to me that the burden of proof of innocence in the advancement of misogyny belongs to the science enthusiasts, not to the victims of these genderized meanings.

THE SOCIAL CONSTRUCTION OF HUMAN SEXUALITY

So far has popular sensibility come in just a few decades that eighteenth- and nineteenth-century assumptions about sex difference may initially seem as alien and incomprehensible to most readers as the beliefs of a medieval peasant or of the original "woman the gatherer." Lillian Faderman writes that what to modern eyes would be regarded as relatively amateurish efforts at cross-dressing, at transvestism, were rarely detected prior to the popularization of Freudian theories and androgynous clothing styles. Dress was taken as a clear indicator of sex: "If a woman craved freedom in a pre-unisex fashion era, when people believed that one's garments unquestionably told one's sex and there was no need to scrutinize facial features and muscle structure to discern gender, she might attempt to pass as a man."[19] How unimaginable that clothes could be an unambiguous indicator of sex! (Why we should be so preoccupied—except for a few hours once in a while—with the sex of the friends and strangers with whom we find ourselves in interaction is another and mysterious matter.)

Simone de Beauvoir's analysis in *The Second Sex* was one important stimulant to the emergence of current theories of the social construction of perceived sex differences, sexuality, and gender. Other contributions to this new consciousness have been made by biological, historical, anthropological, and psychological studies of changes and variety in meanings of masculinity and femininity. The research is very recent, and thus it is hard for most of us to grasp that very little in the forms of our own and others' gender and sexual identities, practices, or desires is given by nature. It is easier to understand the part played by the scientific world view and the particular sciences in shaping both sex and gender—the sex/gender system—if we can begin to grasp the innate plasticity of both sex and gender for members of our species.

[19]Faderman (1981, 48).

Furthermore, understanding the plasticity of the sex/gender system makes more imaginable the elimination of gender that is apparently required if even the affirmative action goals for women in science are to succeed. For a variety of reasons, then, we need to grasp the challenge to biological determinism presented by the social constructionists. Biological determinism is not the only reasonable response to the erosion of borders between nature and culture.[20] Women's place in the sex/gender system is socially constructed, but so is men's. Biological, historical, anthropological, and psychological studies all provide evidence for these claims.

To begin with biology, sex researchers argue that human sexuality is fundamentally extremely plastic, not rigidly controlled by genetic or hormonal patterning.[21] Human infants are born bisexual or "polymorphously perverse," in Freud's phrase. Of course, males inseminate and females incubate and lactate; the male and female development processes that account for this reproductive difference are defined in terms of five biological criteria: genes or chromosomes, hormones, gonads, internal reproductive organs, and external genitalia. But the distance from this biological sex difference to the full-blown construction of gendered and sexual identities, behaviors, roles, and desires in adults is great, and it is evidently traversed entirely by culture. Research on the sexual identity of hermaphrodites, for example, shows complete disjunction between the physiological sex of the hermaphrodite infant and the eventual sex/gender identity adopted by the child. It is parental expectation, not physiological sex, that predicts the adult sex/gender identity for the hermaphroditic infant.[22] And what is true for these cases that come to the attention of scientists because of their abnormality (between 2 and 3 percent of humans are estimated to be hermaphrodites) also appears to be true for the rest of us: social expectation produces sex/gender identities. Furthermore, as indicated earlier, our expectations about biology are shaped by social forces. Research in both biology and the history of biology lead to the inference that the social order creates the biological conceptions that are thought to serve the needs of those holding, aspiring to, or defending power;

[20]Haraway (1985).

[21]See Frank A. Beach, "Evolutionary Changes in the Physiological Control of Mating Behavior in Mammals," *Psychological Review* 54 (1947): 297–313; John Money, "Psychosexual Differentiation," in John Money, ed., *Sex Research: New Developments* (New York: Holt, Rinehart & Winston, 1965), pp. 3–23; John Money and Patricia Tucker, *Sexual Signatures: On Being a Man or a Woman* (Boston: Little, Brown, 1975).

[22]Money, "Psychosexual Differentiation."

and the discipline of biology pays back with interest the support it borrows from the social order.

In addition to Merchant and Jordanova, who focus specifically on the role of the sciences in the historical shifts of meaning and behavior in human sexuality, many other authors have examined such shifts within the more general frameworks of social history. Faderman's book explores the exaltation by men and women—prior to the popularization of Freud and prior to the nineteenth-century women's movement—of what modern eyes would identify as lesbianism. In these earlier cultures, passionate and lifelong friendships between heterosexual women were regarded as normal; indeed, leading masculine authorities considered them moral models for human friendship in general. Such relationships—which may or may not have involved genital sex—began to be labeled lesbian only between 1880 and 1920[23]—years roughly coinciding with the period within which, according to science historian Margaret Rossiter, women waged and lost their fiercest struggles to enter science as equals to men.[24] Faderman's study stands as a challenge to anyone who thinks that heterosexuality refers to the same behaviors and has the same meanings in every time and place. And like Merchant and Jordanova, Faderman shows how men's fear of women's social equality (in this case incited by the nineteenth-century women's movement) and the newly emerging sciences found in each other valuable allies. Psychoanalysis and biomedical research in sex difference gained social legitimacy by defining independent women's support of each other as pathological.

Other historians have scrutinized other aspects of the construction of sex/gender identities and social meanings. Jeffrey Weeks examines the stimulation given to the emergence of self-consciously identified (male) homosexual communities in late nineteenth- and early twentieth-century Europe and America by Freudianism conjoined with repressive legislation against (male) homosexuals. Michel Foucault describes how the masturbating child, the hysterical ("wandering womb," etymologically) woman, and the homosexual male were *created* as objects of scientific scrutiny in the eighteenth and nineteenth centuries. In contrast to the prevailing assumption that the Victorian period was one of unusual repression of discourse about sexuality, Foucault argues that the culture could think of hardly anything else.

[23]Faderman (1981).
[24]Rossiter (1982b).

128

Individuals did engage in what we call masturbating, hysterical, and homosexual behaviors prior to this period, of course, but the creation of types of humans from a subset of their behaviors was a theoretical and political feat of conjoined science and politics, a successful attempt simultaneously to raise the status of science and to develop threat modes of social control for those who did not find congenial the modes of behavior and forms of personal expression desired by an emerging industrial capitalism. Judith Walkowitz describes the creation of a group of people identified as prostitutes in England. Certainly prostitution was not invented in recent history, but the conceptualization of a category of persons as, so to speak, lifelong prostitutes was visibly an invention of that culture. (Walkowitz notes that, ironically, this labeling was aided by the efforts of social reformers to eliminate prostitution.) Many other studies document the changes in the social meanings and behaviors associated with "man" and "woman," "masculine" and "feminine," in Western culture.[25]

The social constructionist strain of recent anthropological literature leaves the impression that there is absolutely nothing—no behavior and no meaning—universally and cross-culturally associated with either masculinity or femininity. What is considered masculine in some societies is considered feminine or gender-neutral in others and vice versa; the only constant appears to be the importance of the dichotomy itself. Two collections of papers in particular take on and explore further the claim originally advanced by Sherry Ortner that in all societies masculinity is associated with culture and femininity with nature—the meanings so evident in the Western societies examined by Merchant and Jordanova.[26] These studies suggest that the nature/culture dichotomy itself, as well as both the particular way the dichotomy is drawn in our society and its gender meanings, are modern and Western. Modern Western meanings of sex/gender and the nature/culture dichotomy have shaped each other. Thus we should be suspicious of cross-cultural generalizations on the basis of what these differences mean and refer to in our society.

On the one hand, the effect of these studies is to challenge the universality of the particular dichotomized set of social behaviors and meanings associated wtih masculinity and femininity in Western culture. For instance, in feminist writings, the concept of unchanging,

[25]Weeks (1981); Foucault (1980); Walkowitz (1983).
[26]Ortner (1974); MacCormack and Strathern (1980); Ortner and Whitehead (1981).

129

universal, "absolute patriarchy" cannot account for the richness and variety of ways in which various cultures work out sex/gender identities or the practices and meanings of social relations between the genders. Furthermore, the very sex/gender dichotomy so central to feminist thinking appears to replicate the nature/culture dualism. Our own analytical categories are probably fatally tainted with echoes and mirror images of the concepts and theories we criticize. On the other hand, there is no society examined by these anthropologists where sex/gender difference is not important. A small but articulate loyal opposition of feminist anthropologists argue that there have been, in some times and places, gender arrangements that were (or are) egalitarian because they were constructed out of gender complementarity instead of gender opposition.[27] But a focus on gender complementarity is still a focus on gender difference. Furthermore, even if these anthropologists are right, masculine dominance appears to be the rule which is at best proved by these possible exceptions.

At the more speculative end of anthropological examination of gender variation, several papers try to reconstruct the initial invention of masculine-dominant gender and sexuality at the dawn of history. Gayle Rubin's widely discussed work provides a feminist reinterpretation of the conjunction of Lévi-Strauss's analysis of the nature of kinship and a Lacanian reading of Freud's analysis of the creation of gender in individuals. Rubin argues that compulsory heterosexuality, marriage, and the division of labor by gender are the causal roots of masculine dominance. Salvatore Cucchiari reaches even deeper into fragile evidence about the origins of human cultures to challenge the assumptions made by Lévi-Strauss and the feminist anthropologists—such as Rubin, Ortner, and Michelle Rosaldo—who draw on him. Cucchiari uses cave paintings as evidence for the discovery/invention of biological sex difference, exclusively female maternity, and eventual masculine domination as objects of human observation and significance *within*—not prior to—distinctively human history. In contrast to Lévi-Strauss and later feminist theorists, Cucchiari argues that the focus on gender dichotomizing did not appear simultaneously with distinctively human culture; it appeared after the invention of tools and language as the unfortunate resolution of a tension between an initially unitary world

[27]E.g., Leacock (1982), and Jane F. Collier and Michelle Z. Rosaldo, "Politics and Gender in Simple Societies," in Ortner and Whitehead (1981).

of human meaning and behavior and the slowly emerging awareness that not everyone could give birth to infants.

Anthropologists have long made us aware that not every culture even today understands paternity: the "hypothesis" that males have something to do with conception is a theoretical achievement. Our culture is so obsessed with sex and gender difference that it is almost impossible for us to imagine a social world in which people did not notice genital difference, or, therefore, that only females give birth to infants. Perhaps appreciation of the discovery of paternity as a great early theoretical achievement will make more plausible Cucchiari's proposal that exclusive female maternity and indeed, reproductive sex difference itself may not have always been commonsense observational givens for humans.[28]

Finally, psychological studies owing debts to Freud have examined the social construction of sexuality and gender in individuals and groups. One stream of this literature has been particularly influential in American feminist writings about science because it shows how women and men come to hold gender-specific models of the self, others, and nature. This is the "object-relations" theory originally developed by D. W. Winnicott, Margaret Mahler, Harry Guntrip, and others and interpreted within a feminist framework by Nancy Chodorow, Dorothy Dinnerstein, and Jane Flax.[29] It is so named because it describes the social/physical mechanism through which adult men and women come to model—to objectify—themselves and their relations to the world in very different ways. In cultures where most child care is performed by women, both male and female infants must individuate themselves against only women. This struggle creates different models of the self and its relation to others for those who are becoming girls and boys. Because the creation of gender in the individual occurs simultaneously within the transformation of a neonate of our species into a social person, our social identities as distinctive human beings are inseparable from our sexual identities as female and male or our gender identities as feminine and masculine.

[28]Rubin (1975); Cucchiari (1981).

[29]D. W. Winnicott, *The Maturational Processes and the Facilitating Environment* (New York: International University Presses, 1965); Margaret Mahler, Fred Pine, and Anni Bergman, *The Psychological Birth of the Human Infant* (New York: Basic Books, 1975); Harry Guntrip, *Personality Structure and Human Interaction* (New York: International Universities Press, 1961); Chodorow (1978); Dinnerstein (1976); Flax (1978; 1983). See Harding (1980; 1981) for my discussion of feminist object-relations theory, upon which this section draws.

Object-relations theorists point out that the biological birth of the human infant is a different process from the psychological birth of the social person. The former is an event of short duration (nine months or a few hours, depending on how one conceptualizes it), relatively uninfluenced by social variables. The latter is a process in which the fundamental stages take about three years, and which is strongly influenced by its social environment. Psychological birth is the first distinctively human labor. Far from being a passive recipient of external stimuli, the infant struggles to emerge from its initial oneness with the psycho/physical environment of its caretakers—which, in societies with our asymetrically valued division of labor by sex/gender, is the mother-world. This first social labor of the infant is extremely difficult and painful because the infant wants to remain in, or return to, that oneness with the mother-world but also to become a separate person. And the infant is particularly vulnerable to its caretakers' projects; it is physically and emotionally dependent on its caretakers for the necessities of life and for recognition of its struggles.

For children of both sexes, the world from which they must differentiate themselves, and against which they discover/create their own autonomous identity, is in one sense the same world: the mother-world. But in another sense it is a very different world for male and female infants: gender-differentiated experiential worlds begin at birth. These theorists argue that the masculine personality develops through separation and individuation from a kind of person whom biologically he cannot become and whom he must exercise will and control not to become socially—a devalued woman. Her body, experientially embedded as it is for him in the whole mother-world, becomes the first model for the bodies and worlds of others—of persons who are perceived as unlike himself and against which, at risk of losing his self-identity, he must create and maintain a strong sense of separation and control. His ego boundaries become relatively rigid. In contrast, feminine personality develops through the young female's struggle to separate and individuate from a kind of person whom she will in fact nevertheless become—a devalued woman. Her ego boundaries remain relatively flexible.

Evidently, the mothering received by boys and girls is different. "Mothers tend to experience their daughters as more like and continuous with, themselves. Correspondingly, girls tend to remain part of the dyadic primary mother-child relationship itself. This means that a girl continues to experience herself as involved in issues of merging and separation, and in an attachment characterized by primary identification and the fusion of identification and object choice." In contrast, mothers experience a son as a masculine opposite; as a result, "boys are more likely to have been pushed out of the preoedipal relationship and to have had to curtail their primary love and sense of empathetic tie with their mother." Consequently, boys' development entails "a more emphatic individuation and a more defensive firming of ego boundaries." For boys, but not for girls, "issues of differentiation have become intertwined with sexual issues."[30]

According to these analyses, masculinity is defined through the achievement of separation, while femininity is defined through the maintenance of attachment. This causes masculine gender identity to be threatened by intimacy or by close identification with particular others' needs and interests, whereas feminine gender identity will be threatened by separation from others and by too little identity with others' needs and interests. For boys, the project of elaborating rules for social interaction helps to ensure smoothly functioning relationships without requiring personal involvement in sustaining either the relationships or the others in them.

Dinnerstein suggests that ecological disaster and the taste for militarism have roots in this masculine gendering process. Flax has pointed to central intellectual structures in the thinking of Plato, Descartes, Hobbes, and Rousseau that appear to be expressions of the "normal" arrested social development of masculine toddlers. And Keller gives a brief account of the relevance of object-relations theory to feminist concerns about science. Other theorists provide analyses that have implications for feminist criticisms of the androcentrism of science: Carol Gilligan's book on moral development theory uses the feminist object-relations analysis to explain the gender differences she found in her study of American children's and adults' conceptions of what constitutes a moral problem and of how moral problems should be resolved.[31] The rules of scientific inquiry are moral norms no less than the principles we adopt for decision-making in social life more generally; thus we should not be surprised to find in scientific method and scientific rationality masculine conceptions of the relations that *should* exist between self, others, and nature.[32] In another study, Isaac Balbus draws on the object-relations theory to suggest that we should

[30]Chodorow (1978; 166–67).
[31]Dinnerstein (1976); Flax (1983); Keller (1978; 1984); Gilligan (1982).
[32]Harding (1980; 1982).

be able to predict and explain historically and cross-culturally different conceptions of nature and appropriate human relations to nature if we look at cultural differences in child-raising practices.[33]

These feminist rereadings of object-relations theory are not without their limitations or their critics. Nor are they the only rereadings of Freud intended to describe and explain in a feminist way the social construction of gender and sexuality in individuals. Lacanian psychoanalytic theory, which sticks closer to Freud's focus on the oedipal drama than do the object-relations theories, has been a resource for feminists in France and England.

In summary, recent research in biology, history, anthropology, and psychology has converged to make completely implausible the assumptions that human gender and sexuality—identities, behaviors, roles, and desires—are determined by the sex differences necessary for reproduction. De Beauvoir points out that women are made, not born; the subsequent literature shows that not only women but men are socially constructed.

If the masculinity of science expresses not a set of biologically given characteristics of males, but socially constructed identities, practices, and desires; if this masculinism is dangerous and undesirable for men as well as women—are the intellectual, ethical, and political structures of science also dangerous and undesirable?

In this brief review of some of the research contributing to new understandings of the social construction of gender and sexuality, we have seen that many of these studies directly implicate science in particular historical shifts in the meanings and behaviors associated with maleness and femaleness, with masculinity and femininity. Science has usually been allied with new and more powerful definitions of and prescriptions for masculine dominance and androcentrism, and the gender order has often provided support in return for the attempts of emerging sciences to gain social legitimacy. This conjunction of science's role in the social construction of gender and sexuality with a masculine-dominant social order's role in legitimating scientific authority for the purpose of increased social power is the focus of the most radical of the feminist challenges to science.

It is in looking at the relationships in particular modern cultures between individual gender and sexual identities and behaviors, the

[33]Balbus (1982).

134

actual divisions of social labor by sex/gender, and the forms of gender symbolism these cultures favor that we can begin to explain science's deep and complex involvement in advancing an androcentric culture. We can begin to grasp how mystifying is science's claim to be objective, dispassionate, value-neutral, and therefore inherently socially progressive. In the words of Virginia Woolf: "Science it would seem is not sexless; he is a man, a father and infected too."

6 FROM FEMINIST EMPIRICISM
TO FEMINIST STANDPOINT
EPISTEMOLOGIES

The androcentric ideology of contemporary science posits as necessary, and/or as facts, a set of dualisms—culture vs. nature; rational
mind vs. prerational body and irrational emotions and values; objectivity vs. subjectivity; public vs. private—and then links men and
masculinity to the former and women and femininity to the latter in
each dichotomy. Feminist critics have argued that such dichotomizing
constitutes an ideology in the strong sense of the term: in contrast to
merely value-laden false beliefs that have no social power, these beliefs
structure the policies and practices of social institutions, including
science.[1]

Could there be an alternative mode of knowledge-seeking not structured by this set of dualisms? Many feminists have been hesitant to
claim that a specifically feminist science or epistemology is possible—
or at least that we can now envision what such a science and epistemology would look like. Historian of science Donna Haraway believes
that feminists need to consider such questions as these:

> Is there a specifically feminist theory of knowledge growing today which
> is analogous in its implications to theories which are the heritage of Greek
> science and of the scientific revolution of the seventeenth century? Would
> a feminist epistemology informing scientific inquiry be a family member

[1]See the papers in MacCormack and Strathern (1980), which argue that these par-
ticular dualisms are Western and modern. For criticisms, see Fee (1981); Griffin (1978);
Hubbard, Henifin, and Fried (1982); Jordanova (1980); Keller (1984); Harding and
Hintikka (1983); Merchant (1980); Rose (1983); Stehelin (1979).

to existing theories of representation and philosophical realism? Or should feminists adopt a radical form of epistemology that denies the possibility of access to a real world and an objective standpoint? Would feminist standards of knowledge genuinely end the dilemma of the cleavage between subject and object or between noninvasive knowing and prediction and control? Does feminism offer insight into the connections between science and humanism? Do feminists have anything new to say about the vexed relations of knowledge and power? Would feminist authority and power to name give the world a new identity, a new story?[2]

AMBIVALENCE AND TRANSITION

Haraway is skeptical that feminist theory (at least in its 1981 form, when she formulated these challenges) can provide the answers. Her questions were prompted by an ambivalence within feminist thinking about science that is still problematic. One form this ambivalence takes is the appeal to Kuhnian arguments: men see the world in one way, women in another; on what possible grounds other than gender loyalties can we decide between these conflicting accounts? For example, to some observers this appears to be the state of the "man-the-hunter" vs. "woman-the-gatherer" hypotheses we examined in Chapter 4.[3] But feminists who deny the possibility of access to a real world and an objective standpoint appear to cut off the possibility of a degendered science at all. Of course, such relativist accounts are responding to the well-founded belief that philosophical and scientific appeals to objectivity and value-free inquiry have often merely provided covers for the refusal to scrutinize critically the social values and projects that have played an important role in the history of science and its intellectual structures. But does our recognition of the fact that science has always been a social product—that its projects and claims to knowledge bear the fingerprints of its human producers—require the exaltation of relativist subjectivity on the part of feminism?

Haraway is certainly right to question whether the feminist critique of "objectivism" (the assumption that objectivity must always be satisfied by value-neutrality) forces us to "subjectivism," to relativism (the assumption that no value-directed inquiries can be objective and therefore all are equally justifiable). Does not this subjectivism leave un-

[2]Haraway (1981, 470).
[3]Longino and Doell (1983). But see Caulfield (1985) and Zihlman (1985) for different assessments of the epistemological and political status of feminist contributions to evolutionary theory.

challenged far too much of the opposition between facts and values, "pure science" and moral/political society, claimed by the science we have? After all, the science we have is highly incorporated into the projects of a bourgeois, racist, and masculine-dominant state, military, and industrial complex. Is "different strokes for different folks" the most defensible and powerful response that can be made to the life-threatening projects supported by the science we have?

The leap to relativism also misgrasps feminist projects. The leading feminist theorists do not try to substitute one set of gender loyalties for the other—"woman-centered" for "man-centered" hypotheses. They try instead, to arrive at hypotheses that are free of gender loyalties. It is true that first we often have to formulate a "woman-centered" hypothesis in order even to comprehend a gender-free one. But the goal of feminist knowledge-seeking is to achieve theories that accurately represent women's activities as fully social, and social relations between the genders as a real—an explanatorily important—component in human history. There is nothing "subjective" about such a project, unless one thinks only visions distorted by gendered desires could imagine women to be fully social and gender relations to be real explanatory variables. From the perspective of feminist theory and research, it is *traditional* thought that is subjective in its distortion by androcentrism—a claim that feminists are willing to defend on traditional objectivist grounds.

The ambivalence also appears when feminists appeal to scientific "facts" to refute sexist claims to provide scientific "facts," while simultaneously denying possibility of perceiving any reality "out there" apart from socially constructed languages and belief systems. Haraway points out that this ambivalent stance is often taken by the same feminist scientists who have provided the most powerful criticisms of "objectivism." How can we appeal to our own scientific research in support of alternative explanations of the natural and social world that are "less false" or "closer to the truth," and at the same time question the grounds for taking scientific facts and their explanations to be the reasonable end of justificatory arguments? As Longino and Doell phrased the issue, how can we simultaneously question both "bad science" and "science-as-usual"?

Another problem that may have motivated Haraway's questions is raised by Elizabeth Fee. Should we look for an alternative science in laboratory procedures, in the methods and modes of reasoning that feminist scientists use? As some hostile skeptics are wont to ask: "Does feminism have an alternative to deduction and induction? To observation and experiment? If not, what could be meant by a feminist science?" We considered in Chapter 2 the distorted conception of science motivating these kinds of questions. Arguing that "at this historical moment, what we are developing is not a feminist science, but a feminist critique of existing science," Fee proposes that we must first bring about a feminist society before we can even begin to imagine a feminist science. "We can expect a sexist society to develop a sexist science; equally we can expect a feminist society to develop a feminist science. For us to imagine a feminist science in a feminist society is rather like asking a medieval peasant to imagine the theory of genetics or the production of a space capsule; our images are, at best, likely to be sketchy and unsubstantial."[4] Fee is certainly right to stress the importance of feminist practice to feminist theory, and the consequent limitations on our ability to imagine the intellectual structures of a world we do not yet have. But must a feminist program for new understanding of knowledge-seeking remain on the back burner until we achieve a feminist society? Does theory come entirely after practice? Or does it emerge as an ongoing process from the struggles in which we engage to bring about a feminist society? And will the fundamental novelties of a feminist science be found in its substantive theories and technologies, or in its epistemology—its theory of the possible and desirable relationships between "human nature" and the world we would understand—or, perhaps, in the fit between the two? (How would we answer these questions about modern science itself?)

Some theorists have argued that forerunners or hints of a feminist science can be detected in the alternative practices of present women scientists.[5] It is becoming perfectly clear that many women conceptualize interactions with other people and nature differently than do most Western men, as the feminist object-relations studies reviewed in Chapter 5 indicated. But I think it is a mistake to search through existing or past practices of individual women scientists for the broad outlines of a feminist science. That would be like looking for a vision of the scientific world view in the imaginations not perhaps of Fee's medieval peasants but rather of early Renaissance artisans and the like, whose new kind of labor made possible the ensuing widespread appreciation of the virtues of experimental observation.[6]

[4]Fee (1981, 22).
[5]E.g., Merchant (1980, ch. 11); Keller (1983); Rose (1983).
[6]See Zilsel (1942), and my discussion in Chapter 9.

Women scientists do violate the division of labor by gender which restricts women to domestic work or low-status wage labor. But how alternative can the practices be of isolated individuals who have somehow managed to bridge this division of labor and social identity? The research agendas of the natural sciences are set in international circles—not by isolated researchers in local laboratories. The existing social structure of science (reviewed in Chapter 3) is an obstacle to the expression within science of whatever unique talents and abilities individual women scientists may have. Furthermore, is a feminist science simply the collection of women scientists' alternative concepts and practices, isolated from any direction by the shifting and diverse understandings and goals of feminist theory and the women's movement? Can a science grounded in women's *identities as gendered* be a sound grounding for a *feminist* science?

To locate the possible directions within which a feminist science could emerge, we should look instead to the distinctive theories of knowledge already being developed. What we think of today as "scientific method" took centuries to develop. Only the broadest generalities about procedures of inquiry and their justificatory strategies can link Galileo's "method" with the methods used today by high-energy physicists or by geneticists. (And as we saw in Chapter 2, much of what we think of as scientific method does not in fact distinguish scientific activities from others we do not call scientific—an issue that has preoccupied much of the philosophical post-Kuhnian discourse.) But some of the proposals about knowers, the world to be known, and the process of coming to know that distinguish modern from medieval theories of knowledge were already clearly detectable in the thinking of Galileo and his peers. Similarly, feminist theoreticians have already proposed concepts of knowers, the world to be known, and the process of knowing that distinguish feminist theories of knowledge from the dominant Western views of the last few centuries. It is these alternative feminist theories of knowledge that already implicitly or explicitly direct many feminist inquiry practices.

The questions we recognize as epistemological originated in their modern form as a "meditation" upon the implications of the emergence of modern science itself. Descartes, Locke, Hume, and Kant were trying to make sense of the kind of knowledge-seeking exemplified by Copernicus, Galileo, and Newton. The creators of modern epistemologies were meditating upon what they understood to be a science created by individual "craft-laborers." Their perceptions of the nature and activities of what they took to be the individual, "disembodied," but human mind, beholden to no social commitments but the willful search for clear and certain truth, remain the foundations from which the questions we recognize as epistemological arise. Once we stop thinking of modern Western epistemologies as a set of philosophical givens, we can begin to examine them instead as historical justificatory strategies—as culturally specific modes of constructing and exploiting cultural meanings in support of new kinds of knowledge claims. After all, the legitimacy of the theological justifications once presented for scientific (and mathematical) claims and practices was eventually undercut by the claims and practices of modern science; the scientific claims and practices became more intuitively acceptable than the theologies invoked to justify them.

Similarly, I shall argue that the substance of feminist claims and practices can be used to undercut the legitimacy of the modernist epistemologies, which explicitly ignore gender while implicitly exploiting distinctively masculine meanings of knowledge-seeking. Gender-sensitive revisions of modernist epistemologies have provided the main justificatory resources for feminism—a situation only now coming to be fully recognized by feminist theorists, though forerunners of such recognition can be seen in the ambivalences we have noted. Thus I propose that we think of feminist epistemologies as still transitional meditations upon the substance of feminist claims and practices. In short, we should expect, and perhaps even cherish, such ambivalences and contradictions. In this sense, Fee is right: we will have a feminist science fully coherent with its epistemological strategies only when we have a feminist society.

In this chapter and the next I want to examine the feminist standpoint epistemologies we previewed in Chapter 1, identify some challenges to these epistemologies, and explore the motivation toward feminist postmodernism that such challenges create.

THE FEMINIST STANDPOINT EPISTEMOLOGIES

The feminist standpoint epistemologies ground a distinctive feminist science in a theory of gendered activity and social experience. They simultaneously privilege women or feminists (the accounts vary) epistemically and yet also claim to overcome the dichotomizing that is characteristic of the Enlightenment/bourgeois world view and its sci-

141

ence.[7] It is useful to think of the standpoint epistemologies, like the appeals to feminist empiricism, as "successor science" projects: in significant ways, they aim to reconstruct the original goals of modern science. In contrast, feminist postmodernism more directly challenges those goals (though there are postmodernist strains even in these standpoint writings).

An observer of these arguments can pick out five different though related reasons that they offer to explain why inquiry from a feminist perspective can provide understandings of nature and social life that are not possible from the perspective of men's distinctive activity and experience. I shall identify each of these reasons in the writing of one theorist who has emphasized this particular aspect of the gendered division of activity, though most of these theorists recognize more than one. Whatever their differences, I think the accounts should be undestood as fundamentally complementary, not competing.

The Unity of Hand, Brain and Heart in Craft Labor.

Hilary Rose's "feminist epistemology for the natural sciences" is grounded in a post-Marxist analysis of the effects of gendered divisions of activity upon intellectual structures.[8] In two recent papers, she has developed the argument that it is in the thinking and practices of women scientists whose inquiry modes are still characteristically "craft labor," rather than the "industrialized labor" within which most scientific inquiry is done, that we can detect the outlines of a distinctively feminist theory of knowledge. Its distinctiveness is to be found in the way its concepts of the knower, the world to be known, and processes of coming to know reflect the unification of manual, mental, and emotional ("hand, brain, and heart") activity characteristic of women's work more generally. This epistemology not only stands in opposition to the Cartesian dualisms—intellect vs. body, and both vs. feeling and emotion—that underlie Enlightenment and even Marxist visions of science but also grounds the possibility of a "more complete materialism, a truer knowledge" than that provided by either paternal discourse (1984, 49). The need for such a feminist science "is increasingly acute," for "bringing caring labor and the knowledge that stems from

---

[7]The offensively dichotomized categories of labor vs. leisure, which appear in the parental Enlightenment/bourgeois and Marxist theories, are themselves the target of criticism in the standpoint epistemologies; it is a theory of human *activity* and social experience they are proposing.

[8]Rose (1983; 1984). Subsequent page references to these papers appear in the text.

participation in it to the analysis becomes critical for a transformative program equally within science and within society" if we are to avoid the nuclear annihilation and deepening social misery increasingly possible otherwise (1983, 89).

Rose starts by analyzing the insights of post-Marxist thinking upon which feminists can build. Sohn-Rethel saw that it was the separation of manual from mental labor in capitalist production that resulted in the mystifying abstractions of bourgeois science.[9] But social relations include far more than the mere production of commodities where mental and manual labor are assigned to different classes of people. Like Marx, Sohn-Rethel failed to ask about the effect on science of assigning *caring* labor exclusively to women.[10] Rose argues that in this respect, post-Marxists such as Sohn-Rethel are indistinguishable from the sociobiological theorists to whom they are vehemently opposed; they tacitly endorse the "far-from-emancipatory program of sociobiology, which argues that woman's destiny is in her genes." Feminists must explain the relationship between women's unpaid and paid labor to show that women's caring skills have a social genesis, not a natural one, and that they "are extracted from them by men primarily within the home but also in the work place" (1983, 83–84).

Rose goes on to analyze the relationship of the conditions of women's activities within science with those in domestic life, and the possibilities created by these kinds of activities for women to occupy an advantaged standpoint as producers of less distorted and more comprehensive scientific claims. A feminist epistemology cannot originate in meditations upon what women do in laboratories, since the women there are forced to deny that they are women in order to survive, yet are still "by and large shut out of the production system of scientific knowledge, with its ideological power to define what is and what is not objective knowledge" (1983, 88). They are prohibited from becoming (masculine) scientific knowers and also from admitting to being what they are primarily perceived as being: women.[11]

In her earlier paper, Rose argues that a feminist epistemology must be grounded in the practices of the women's movement. In its consideration of such biological and medical issues as menstruation, abortion, and self-examination and self–health care, the women's movement fuses "subjective and objective knowledge in such a way as to make new

---

[9]Sohn-Rethel (1978).
[10]Hartsock (1983b; 1984) also raises this criticism about Sohn-Rethel.
[11]Cf. the discussion of this dilemma in Stehelin (1979).

knowledge." "Cartesian dualism, biological determinism, and social constructionism fade when faced with the necessity of integrating and interpreting the personal experience of [menstrual] bleeding, pain, and tension," Rose declares. "Working from the experience of the specific oppression of women fuses the personal, the social, and the biological." Thus a feminist epistemology for the natural sciences will emerge from the interplay between "new organizational forms" and new projects (1983, 88–89). The organizational forms of the women's movement, unlike those of capitalist production relations and its science, resist dividing mental, manual, and caring activity among different classes of persons. And its project is to provide the knowledge women need to understand and manage our own bodies: subject and object of inquiry are one. Belief emerging from this unified activity in the service of self-knowledge is more adequate than that emerging from activity that is divided and that is performed for the purposes of monopolizing profit and social control.

This first paper left a gap between the kind of knowledge/power relations possible in a science grounded in women's understandings of our own bodies and the kind needed if a feminist science is to develop sufficient muscle to replace the physics, chemistry, biology, and social sciences we have. In the later paper, Rose inches across this gap by expanding the domain in which she thinks we can identify the origins of a distinctive feminist epistemology. The origins of an epistemology which holds that appeals to the subjective are legitimate, that intellectual and emotional domains must be united, that the domination of reductionism and linearity must be replaced by the harmony of holism and complexity, can be detected in what Foucault would call "subjugated knowledges"—submerged understandings within the history of science (1984, 49).

Rose has in mind here the ecological concerns reported and elaborated by Carolyn Merchant and evident in Rachel Carson's work, and the calls for moving beyond reductionism toward a holistic "feminization of science" evident in writers such as David Bohm and Fritjof Capra.[12] She might also have cited here Joseph Needham's romantic idealization of Chinese science as more feminized than Western sci-

---

[12]Merchant (1980); Rachel Carson, *Silent Spring* (New York: Fawcett, 1978, originally published in 1962); David Bohm, *Wholeness and the Implicate Order* (Boston: Routledge & Kegan Paul, 1980); Fritjof Capra, *The Tao of Physics* (New York: Random House, 1975).

ence.[13] And then we would have to think about the contradictions between China's history of a "feminized science" and the far from emancipatory history of Chinese misogyny. This raises the troublesome issue of the conflation of gender dichotomies as a metaphor for other dichotomies (gender symbolism) with explanations that treat social relations between the sexes as a causal influence on history—a point to be pursued later. Furthermore, this line of thought leads directly toward feminist distrust of men's conceptions of the androgyny men desire for themselves. When men want androgyny, they usually intend to appropriate selectively parts of "the feminine" for their projects, while leaving the lot of real women unchanged.[14]

Within recent scientific research by women in biology, psychology, and anthropology—areas where "craft" forms of scientific inquiry are still possible, in contrast to the "industrial" forms confronting women in masculine-dominated labs—Rose detects the most significant advances toward "a more complete materialism, a truer knowledge." In all of these areas, feminist thinking has produced a new comprehension of the relationships between organisms, and between organisms and their environment. The organism is conceptualized "not in terms of the Darwinian metaphor, as the passive object of selection by an indifferent environment, but as [an] active participant, a subject in the determination of its own future" (1984, 51). (Keller has argued that Barbara McClintock's work provides a paradigm of this kind of alternative to the "master theory" of Darwinian biology.[15])

Thus Rose proposes that the grounds for a distinctive feminist science and epistemology are to be found in the social practices and conceptual schemes of feminists (or women inquirers) in craft-organized areas of inquiry. There women's socially created conceptions of nature and social relations can produce new understandings that carry emancipatory possibilities for the species. These conceptions are not necessarily original to women scientists: hints of them can be detected in the "subjugated knowledges" in the history of science. However, we can here hazard an observation Rose does not make: where these notions neither originate in nor give expression to any distinctive social/

[13]Needham (1976).

[14]See Bloch and Bloch (1980) on the deradicalization of the thought of Rousseau and other French thinkers that occurred once they recognized that the logic of their radical arguments was about to lead them directly to the conclusion that "the good" which should direct the social order was identical to what, in fact, women do.

[15]Keller (1983).

political experience, they are fated to remain mere intellectual curiosities—like the ancient Greek ideas about atoms—awaiting their "social birth" within the scientific enterprise at the hands of a group which needs such conceptions in order to project onto nature its destiny within the social order. One cannot help noticing that the notion of organisms as active participants in the determination of their own futures "discovers" in "nature" the very relationship that feminist theory claims has been permitted only to (dominant group) men but *should* exist as well for women, who are also history-making social beings. Men have actively advanced their own futures within masculine domination; women, too, could actively participate in the design of their futures within a degendered social order.

Whether or not Rose would agree to this conclusion, she does argue that the origins of a feminist epistemology for a successor science are to be found in the conceptions of the knower, the processes of knowing, and the world to be known which are evident in this substantive scientific research. The substantive claims of this research are thus to be justified in terms of women's different activities and social experiences created in the gendered division of labor/activity. As I shall ask of each of these standpoint theorists, does this epistemology still retain too much of the Enlightenment vision?

Women's Subjugated Activity: Sensuous, Concrete, Relational.
Like Rose, political theorist Nancy Hartsock locates the epistemological foundations for a feminist successor science in a post-Marxist theory of labor (activity) and its effects upon mental life. For Hartsock, too, Sohn-Rethel provides important clues. But Hartsock begins with Marx's metatheory, his "proposal that a correct vision of class society is available from only one of the two major class positions in capitalist society."[16] By starting from the lived realities of women's lives, we can identify the grounding for a theory of knowledge that should be the successor to both Enlightenment and Marxist epistemologies. For Hartsock as for Rose, it is in the gendered division of labor that one can discover both the reason for the greater adequacy of feminist knowledge claims, and the root from which a full-fledged successor to Enlightenment science can grow. However, the feminist successor science will be anti-Cartesian, for it transcends and thus stands in opposition

---

[16]Hartsock (1983b, 284). This paper also appears as ch. 10 in Hartsock (1984). Page numbers in the text refer to the 1983 version.

to the dichotomies of thought and practice created by divisions between mental and manual labor, though in a way different from that which Rose identifies.

Women's activity consists in "sensuous human activity, practice." Women's activity is institutionalized in two kinds of contributions—to "subsistence" and to child-rearing. In subsistence activities, contributions to producing the food, clothing, and shelter necessary for the survival of the species,

> the activity of a woman in the home as well as the work she does for wages keeps her continually in contact with a world of qualities and change. Her immersion in the world of use—in concrete, many-qualitied, changing material processes—is more complete than [a man's]. And if life itself consists of sensuous activity, the vantage point available to women on the basis of the contribution to subsistence represents an intensification and deepening of the materialist world view and consciousness available to the producers of commodities in capitalism, an intensification of class consciousness. [p. 292]

However, it is in examining the conditions of women's activities in child care that the inadequacy of the Marxist analysis appears most clearly. "Women also produce/reproduce men (and other women) on both a daily and a long-term basis. This aspect of women's 'production' exposes the deep inadequacies of the concept of production as a description of women's activity. One does not (cannot) produce another human being in anything like the way one produces an object such as a chair. . . . Helping another to develop, the gradual relinquishing of control, the experience of the human limits of one's action" are fundamental characteristics of the child care assigned exclusively to women. "The female experience in reproduction represents a unity with nature which goes beyond the proletarian experience of interchange with nature" (p. 293).

Furthermore, Hartsock draws on the feminist object-relations theory of Jane Flax and Nancy Chodorow to show that women are "made, not born" in such a way as to define and experience themselves concretely and relationally.[17] In contrast, newborn males are turned into men who define and experience themselves abstractly and as fundamentally isolated from other people and nature. Not-yet-gendered newborn males and females are shaped into the kinds of personalities

[17]Flax (1983); Chodorow (1978).

147

who will want to perform characteristic masculine and feminine activities. The consequences that object-relations theorists describe are just what Hartsock finds when she examines the adult division of labor by gender: relational femininity vs. abstract masculinity. Both the epistemology and the society constructed by "men suffering from the effects of abstract masculinity" emphasize "the separation and opposition of social and natural worlds, of abstract and concrete, of permanence and change"—the same oppositions as those stressed in the Marxist analysis of bourgeois labor. Thus the true counter to the bourgeois subjugations and mystifications is not to be found in a science grounded in proletarian experience, for this is fundamentally still a form of men's experience; it is instead to be found in a science grounded in women's experience, for only there can these separations and oppositions find no home (pp. 294–98).

The conditions under which women contribute to social life must be generalized for all humans if an effective opposition to androcentric and bourgeois political life and science/epistemology is to be created. Politically, this will lead to a society no longer structured by masculinist oppositions in either their bourgeois or proletarian forms; epistemologically, it will lead to a science that will both direct and be directed by the political struggle for that society.

A feminist epistemological standpoint is an interested social location ("interested" in the sense of "engaged," not "biased"), the conditions for which bestow upon its occupants scientific and epistemic advantage. The subjugation of women's sensuous, concrete, relational activity permits women to grasp aspects of nature and social life that are not accessible to inquiries grounded in men's characteristic activities. The vision based on men's activities is both partial and perverse—"perverse" because it systematically reverses the proper order of things: it substitutes abstract for concrete reality; for example, it makes death-risking rather than the reproduction of our species form of life the paradigmatically human act. Even early feminists such as Simone de Beauvoir think within abstract masculinity: "It is not in giving life but in risking life that man is raised above the animal: that is why superiority has been accorded in humanity not to the sex that brings forth but to that which kills."[18]

Moreover, men's vision is not simply false, for the ruling group can make their false vision become apparently true: "Men's power to struc-

---

[18]Simone de Beauvoir (1953, 58), cited in Hartsock (1983, 301).

148

ture social relations in their own image means that women, too, must participate in social relations which manifest and express abstract masculinity" (p. 302). The array of legal and social restrictions on women's participation in public life makes women's characteristic activities appear to both men and women as merely natural, as merely continuous with the activities of female termites or apes (as the sociobiologists would have it), and thus as suitable objects of men's manipulations of whatever they perceive as purely natural. The restriction of formal and informal educational opportunities for women makes women appear incapable of understanding the world within which men move, and as appropriately forced to deal with that world in men's terms.

The vision available to women "must be struggled for and represents an achievement which requires both science to see beneath the surface of the social relations in which all are forced to participate, and the education which can only grow from struggle to change those relations" (p. 285). The adoption of this standpoint is fundamentally a moral and political act of commitment to understanding the world from the perspective of the socially subjugated. It constitutes not a switch of epistemological and political commitments from one gender to the other but a commitment to the transcendence of gender through its elimination. Such a commitment is social and political, not merely intellectual.

Hartsock is arguing that divisions of labor more intensive than those Marx identified create dominating political power and ally perverse knowledge claims with the perversity of dominating power. Therefore, a science generated out of a transcendence, a transformation, of these divisions and their corresponding dualisms will be a powerful force for the elimination of power. In an earlier paper, Hartsock argued that the concept of power central to the history of political theory is only one available concept. Against power as domination *over* others, feminist thinking and organizational practices express the possibility of power as the provision of energy *to* others as well as self, and of reciprocal empowerment.[19] I think this second notion of power and the kind of knowledge that could be allied with it can remove the apparent paradox from her adoption of both successor science and postmodern tendencies. One can insist on an epistemology-centered philosophy only if the "policing of thought" that epistemology entails is a reciprocal project—with the goal of eliminating the kind of dom-

[19]Hartsock (1974).

149

inating power that makes the policing of thought necessary.[20] That is, such an epistemology would be a transitional project, as we transform ourselves into a culture uncomfortable with domination and thereby into peoples whose thought does not need policing.

Hartsock's grounds for a feminist epistemology are both broader and narrower than Rose's. They are narrower in that it is feminist political struggle and theory ("science")—not simply characteristic women's activities—in which the tendencies toward a specifically feminist epistemology can be detected. Unmediated by feminist struggle and analysis, women's distinctive practices and thinking remain part of the world created by masculine-domination.[21] But her grounds are also broader, for any feminist inquiry that starts from the categories and valuations of women's subsistence and domestic labor and is *interested* (again in the sense of *engaged*) in the struggle for feminist goals provides the grounding for a distinctive epistemology of a successor to Enlightenment science. The women's health movement and the alternative understandings of the relationship between organism and environment that Rose points to would provide significant examples of such inquiries (insofar as they are motivated by the goals of feminist emancipation). But so would any of the natural or social science inquiries that begin by taking women's activities as fully social and try to explain nature and social life for feminist political purposes. There is still a significant gap in Hartsock's account between feminist activity and a science/epistemology robust and politically powerful enough to unseat the Enlightenment vision. But in both its broader and narrower aspects, Hartsock's account inches yet further across the gap by extending the foundation for the successor science to the full array of feminist political and scientific projects and, at least implicitly, to activities in which men as well as women feminists engage.

There is an another important difference in the groundings these two theorists identify for the successor epistemology. Hartsock does not directly focus on the "caring" labor of women, which Rose takes to be the distinctive human activity missing in the Marxist accounts. For Hartsock, the uniqueness of women's labor, in contrast to proletarian labor, is to be found in its more fundamental opposition to the mental/manual dualities that structure masculine/bourgeois thought

---

[20]This critique of epistemology-centered philosophy and its policing of thought is central to the postmodernists. See, e.g., Rorty (1979) and Foucault (1980).

[21]Rose would probably agree with this; many of her other writings would support such an argument. See, e.g., the papers in Rose and Rose (1976).

and activity. For Hartsock, (men's) proletarian labor is transitional between bourgeois/masculine and women's labor, since women's labor is more fundamentally involved with the self-conscious, sensuous processing of our natural/social surroundings in daily life—is the distinctively human activity. For Rose, women's labor is different in kind from (masculine) proletarian/bourgeois labor.

The "Return of the Repressed" in Feminist Theory.

Jane Flax, a political theorist and psychotherapist, explicitly describes the successor science and postmodern tendencies in feminist epistemology as conflicting. In the later of two papers I shall examine, she argues for the postmodern direction to replace the successor science tendency, yet in both papers the two tendencies are linked in a way that evidently appears noncontradictory to her.

In a paper written in 1980, though not published until 1983, Flax calls for a "successor science" project:

> The task of feminist epistemology is to uncover how patriarchy has permeated both our concept of knowledge and the concrete content of bodies of knowledge, even that claiming to be emancipatory. Without adequate knowledge of the world and our history within it (and this includes knowing how to know), we cannot develop a more adequate social practice. A feminist epistemology is thus both an aspect of feminist theory and a preparation for and a central element of a more adequate theory of human nature and politics.[22]

"Feminist philosophy thus represents the return of the repressed, of the exposure of the particular social roots of all apparently abstract and universal knowledge. This work could prepare the ground for a more adequate social theory in which philosophy and empirical knowledge are reunited and mutually enriched" (p. 249).

Flax argues that feminist philosophy should ask the question, "What forms of social relations exist such that certain questions and ways of answering them become constitutive of philosophy?" (p. 248). Here a feminist reading of psychoanalytic object-relations theory (see Chapter 5) becomes a useful philosophic tool; it directs our attention to the distinctively gendered senses of self, others, nature, and relations among the three that are characteristic in cultures where infant care is primarily the responsibility of women. For Flax, what is particularly

---

[22]Flax (1983, 269). Subsequent page references appear in the text.

interesting is the fit between masculine senses of self, others, and nature and the definition of what is problematic in philosophy. From this perspective, "apparently insoluble dilemmas within philosophy are not the product of the immanent structure of the human mind and/or nature but rather reflect distorted or frozen social relations" (p. 248). For men more than for women, the self remains frozen in a defensive infantile need to dominate and/or repress others in order to retain its individual identity. In cultures where primary child care is assigned exclusively to women, male infants will develop unresolvable dilemmas concerning the separation of the infantile self from its first "other" and the establishment of individual identity. These are the very same distinctively masculine dilemmas that preoccupy Western philosophers in whose work they appear as "the human dilemma."

Western philosophy problematizes the relationships between subject and object, mind and body, inner and outer, reason and sense; but these relationships would not need to be problematic for anyone were the core self not always defined exclusively against women.

> In philosophy, being (ontology) has been divorced from knowing (epistemology) and both have been separated from either ethics or politics. These divisions were blessed by Kant and transformed by him into a fundamental principle derived from the structure of mind itself. A consequence of this principle has been the enshrining within mainstream Anglo-American philosophy of a rigid distinction between fact and value which has had the effect of consigning the philosopher to silence on issues of utmost importance to human life. [p. 248]

Were women not exclusively the humans against whom infant males develop their senses of a separate and individuated self, "human knowledge" would not be so preoccupied with infantile separation and individuation dilemmas. "Analysis reveals an arrested stage of human development . . . behind most forms of knowledge and reason. Separation-individuation [of infants from their caretakers] cannot be completed and true reciprocity emerge if the 'other' must be dominated and/or repressed rather than incorporated into the self while simultaneously acknowledging difference" (p. 269). Human knowledge can come to reflect the more adult issues of maximizing reciprocity and appreciating difference only if the first "other" is "incorporated into the self" rather than dominated and/or repressed.

Flax's point is *not* that the Great Men in the history of philosophy

152

would have better spent their time on psychoanalytic couches (had they been available) than in writing philosophy. Nor is it that philosophy is nothing but masculine rationalization of painful infantile experience. Rather, she argues that a feminist exposure of the "normal" relations between infantile gendering processes and adult masculine thought patterns "reveals fundamental limitations in the ability of [men's] philosophy to comprehend women's and children's experiences"; in particular, it reveals the tendency of philosophers to take their own experience as paradigmatically human rather than merely as typically masculine (p. 247). We can move toward a feminist epistemology through exposing the infantile social dilemmas repressed by adult men, the "resolutions" of which reappear in abstract and universalizing form as both the collective motive for and the subject matter of patriarchal epistemology. The feminine dimensions of experience tend to disappear in all thinking within patriarchies. But women's experience cannot, in itself, provide a sufficient ground for theory, for "as the other pole of the dualities it must be incorporated and transcended." Thus an adequate feminist philosophy requires "a revolutionary theory and practice. . . . Nothing less than a new stage of human development is required in which reciprocity can emerge for the first time as the basis of social relations" (p. 270).

In this earlier paper, Flax is arguing that infantile dilemmas are more appropriately resolved, less problematic, for women than for men. This small gap between the genders prefigures a larger gap between the defensive gendered selves produced in patriarchal modes of child rearing and the reciprocal, degendered selves that *could* exist were men as well as women primary caretakers of infants, and women as well as men responsible for public life. The forms and processes of knowing as well as what is known will be different for reciprocal selves than for defensive selves. Truly human knowledge and ways of knowing toward which a feminist epistemology points the way, will be less distorted and more nearly adequate than the knowledge and ways of knowing we now have. And while the concepts of reciprocal knowing must be relational and contextual, and thus will no longer enshrine the dualities of Enlightenment epistemology, it is indeed a successor epistemology toward which feminism moves us all.[23]

[23]Although she stresses here women's less defensive "resolution" of infantile separation and individuation dilemmas, see Flax (1978) for a discussion of those unfortunate residues of the feminine infantile dilemma that create tensions within women and for feminist organizations.

Flax's argument in a paper written four years later contrasts sharply with the foregoing argument. Whereas the earlier paper claims that child-rearing practices leave distinctive marks on philosophers as culturally diverse as Plato, Locke, Hobbes, Kant, Rousseau, and contemporary Anglo-American thinkers, the later one is skeptical that there can be a *single* way that patriarchy has permeated thinking. She finds problematic the notion of "*a* feminist standpoint which is more true than previous (male) ones." She says, "Any feminist standpoint will necessarily be partial. Each person who tries to think from the standpoint of women may illuminate some aspects of the social totality which have been previously suppressed with the dominant view. But none of us can speak for 'woman' because no such person exists except within a specific set of (already gendered) relations—to 'man' and to many concrete and different women."

Here it is feminist theory's affinities with postmodern philosophy that Flax finds most distinctive:

> As a type of post modern philosophy, feminist theory shares with other such modes of thought an uncertainty about the appropriate grounding and methods for explaining and/or interpreting human experience. Contemporary feminists join other post modern philosophers in raising important metatheoretical questions concerning the possible nature and status of theorizing itself. . . . Consensus rules on categorization, appraisal, validity, etc. are lacking.[24]

This affinity is more fundamental, she argues, than feminist attempts at successor science projects: "Despite an understandable attraction to the (apparently) logical, orderly world of the Enlightenment, feminist theory more properly belongs in the terrain of post modern philosophy." And yet the substance of this later paper argues for a particular way of understanding gender that Flax thinks should replace the inadequate and confusing ways it is conceptualized in both traditional and feminist social theory. Gender should be understood as relational; gender relations are not determined by nature but are social relations of domination, and feminist theorists "need to recover and write the histories of women and our activities into the accounts and self-understanding of the whole" of social relations.

On the one hand, in effect Flax has located the feminist successor science tendencies as part of the projects of the defensive self which

[24]Flax (1986, 37).

are most evident in men. She identifies postmodern skepticism about the Enlightenment dualities, which ensure the epistemological "policing of thought," as the entering wedge into projects for the reciprocal self. Overcoming the (distinctively masculine) Enlightenment dualities will be possible for our whole culture only after a "revolution in human development." On the other hand, does not Flax's own account of the distorted and frozen social relations characteristic of masculine-dominant societies suggest both that there is "objective basis for distinguishing between true and false beliefs" and that she is herself committed to this kind of epistemology? Even though any particular historical understanding available to feminists ("a feminist standpoint") is partial, may it not also be "more true than previous (male) ones"?

The Bifurcated Consciousness of Alienated Women Inquirers.
Canadian sociologist of knowledge Dorothy Smith has explored in a series of papers what it would mean to construct a sociology that begins from the "standpoint of women." Though her stated concern is sociology, her arguments are generalizable to inquiry in all the social and natural sciences. In the most recent of these papers, she directly articulates the problem of how to fashion a successor science that will transcend the damaging subject-object, inner-outer, reason-emotion dualities of Enlightenment science. "Here, I am concerned with the problem of methods of thinking which will realize the project of a sociology for women; that is, a sociology which does not transform those it studies into objects but preserves in its analytic procedures the presence of the subject as actor and experiencer. Subject then is that knower whose grasp of the world may be enlarged by the work of the sociologist."[25] Smith thinks that the forms of alienation experienced by women inquirers make it possible to carry out what I have been calling successor science and postmodern projects simultaneously and without contradiction.

Like the other theorists, Smith's epistemology is grounded in a successor to the Marxist theory of labor. (It is perhaps inaccurate to conjoin Flax with the others in this respect, unless we focus on her discussion of the process through which the infant becomes a social person as the first human labor, which is divided, of course, by the gender of the "laboring" infant.) Smith eschews questions of the developmental origins of gender; of the origins in men's infantile experiences of the defensive

[25]Smith (1981, 1). See the discussion of Smith's work in Westkott (1979).

155

abstractions of Western social theory, science, and epistemology; and thus of the reasons why men and women *want* to participate in characteristically masculine and feminine activity. That is, she does not discuss the issue of how initially androgynous infantile "animals" of our species interact with their social/physical environments to become the gendered humans we see around us. Like Rose, she turns to the structure of the workplace for women scientists (sociologists) to locate an enriched notion of the material conditions that make possible a distinctively feminist science.

Where Rose focuses on the unity of hand, brain, and heart common to women's characteristic activities, Smith looks at three other shared aspects of women's work. In the first place, it relieves men of the need to take care of their bodies or of the local places where they exist, freeing them to immerse themselves in the world of abstract concepts. Second, the labor of women thereby "articulates," shapes, men's concepts into those of administrative forms of ruling. The more successfully women perform this concrete work (Hartsock's "world of sensuousness, of qualities and change"), the more invisible does their work become to men. Men who are relieved of the need to maintain their own bodies and the local places where they exist can now see as real only what corresponds to their abstracted mental world. Like Hegel's master, to whom the slave's labor appears merely as an extension of his own being and will, men see women's work not as real activity—self-chosen and consciously willed—but only as "natural" activity, as instinctual or emotional labors of love. Women are thus excluded from men's conceptions of culture and its conceptual schemes of "the social," "the historical," "the human." Finally, women's actual experience of their own labor is incomprehensible and inexpressible within the distorted abstractions of men's conceptual schemes. Women are alienated from their own experience, for men's conceptual schemes are also the ruling ones, which then define and categorize women's experience for women. (This is Hartsock's point about ideologies structuring social life for everyone.) For Smith, education for women, for which nineteenth-century feminists struggled, completed the "invasion of women's consciousness" by ruling-class male experts.[26]

[26]Smith (1979, 143). We should note that Smith was writing on these topics earlier than the other theorists I have discussed, though her work did not become widely known in the United States until recently. The aspects of women's labor Smith identifies so clearly and so early also appear to be on the minds of the other theorists, as a perusal of their work will show.

These characteristics of women's activities are a resource that a distinctively feminist science can use. A "line of fault" develops for many women between our own experience of our activity and the categories available to us within which to express our experience: the categories of ruling and of science. The break is intensified for women inquirers. We are first of all women, who—even if single, childless, or with servants—maintain our own bodies and our places of local existence, and usually also the bodies and domestic places of children and men. But when entering the world of science, we are trained to describe and explain social experience within conceptual schemes that cannot recognize the character of this experience. Smith cites the example of time-budget studies, which regard housework as part leisure and part labor—a conceptualization based on men's experience of wage labor for others vs. self-directed activity. But for wives and mothers, housework is neither wage labor nor self-directed activity. An account of housework from "the standpoint of women"—our experience of our lives—rather than in the terms of masculine science would be a quite different account; the voice of the subject of inquiry and the voice of the inquirer would be culturally identifiable.[27] It would be an example of science *for* women rather than *about* women; it would seek to explain/interpret social relations rather than behavior (human "matter in motion"), and do so in a way that makes comprehensible to women the social relations within which their experience occurs.

Smith fuses here what have been incompatible tendencies toward interpretation, explanation, and critical theory in the philosophy of social science. None of these discourses locates "authoritative accounts" in those of the inquirer as an active agent in inquiry. Once Smith puts the authority of the inquirer on the same epistemological plane as the authority of the subjects of inquiry—the women inquirer interpreting, explaining, critically examining women's condition is simultaneously explaining her own condition—then issues of absolutism vs. relativism can no longer be posed. Both absolutism and relativism assume separations between the inquirer and subject of inquiry that are not present when the two share a subjugated social location.[28]

I think Smith is arguing that this kind of science would be "objective," not because it would use the categories available from an "Archimedean," dispassionate, detached "third version" of the conflicting

27 Smith (1979, 154; 1981, 3).
28 Cf. Harding (1980).

perspectives people have on social relations but because it would use the more complete and less distorting categories available from the standpoint of historically locatable subjugated experiences.[29] However, it is difficult to generalize from her explicit assumptions about intepreting/explaining women's world to a feminist science that takes as its project explaining the whole world. She often admonishes the reader that the experience of the subject of inquiry (the experience of the women whose lives the inquirer is explaining) is to be taken as the final authority. But many feminist inquirers take men's experience as well as women's to be inadequately interpreted, explained or criticized within the existing "corpus of knowledge": think of all the recent writing on men's war mentality; of object-relations theory's critical reinterpretation of the masculine experience of gendering; of Smith's own rethinking of men's experiences as sociologists. Yet she does not assign ruling-class men's experience the kind of authority she insists on for women's experience; through all four papers her argument shows why we should regard women's subjugated experience as starting and ending points for inquiry that are epistemologically preferable to men's experience. (Smith's argument here is similar to Hartsock's assertion of the epistemological preferability of the categories of women's activities, and to Flax's focus on feminism as the exposure of what men repress; all three return to Hegel's passage about the master and the slave to make their points.)

Interpreting Smith in this way leaves a few loose ends in her account, but it makes sense of the origins of the scientific authority she clearly intends to give to women as both subjects of inquiry and inquirers. For her, what feminism should distrust is not objectivity or epistemology's policing of thought per se but the particular distorted and ineffectual form of objectivity and epistemology entrenched in Enlightenment science. Like Flax, Smith stresses that there will be many different feminist versions of "reality," for there are many different realities in which women live, but they should all be regarded as producing more complete, less distorting, and less perverse understandings than can a science in alliance with ruling-class masculine activity.

New Persons and the Hidden Hand of History.
Finally, it is historical changes that make possible feminist theory and consequently a feminist science and epistemology, as I have argued

[29]Smith (1981, 6).

elsewhere.[30] Here, too, we can learn from the Marxist analysis. Engels believed that "the great thinkers of the Eighteenth Century could, no more than their predecessors, go beyond the limits imposed upon them by their epoch."[31] He thought that only with the emergence in nineteenth-century industrializing societies of a "conflict between productive forces and modes of production"—a conflict that "exists, in fact, objectively, outside us, independently of the will and actions even of the men that have brought it on"—could the class structure of earlier societies be detected in its fullness for the first time. "Modern socialism is nothing but the reflex, in thought, of this conflict in fact; its ideal reflection in the minds, first, of the class directly suffering under it, the working class."[32]

Similarly, only now can we understand the feminisms of the eighteenth and nineteenth centuries as but "utopian" feminisms.[33] The men and women feminists of those cultures could recognize the misery of women's condition and the unnecessary character of that misery, but both their diagnoses of its causes and their prescriptions for women's emancipation show a failure to grasp the complex and not always obvious mechanisms by which masculine dominance is created and maintained. Liberal feminism, Marxist feminism and perhaps even the more doctrinaire strains of the radical and socialist feminisms of the mid–1970s do not have conceptual schemes rich or flexible enough to capture masculine domination's historical and cultural adaptability, nor its chameleonlike talents for growing within such other cultural hierarchies as classism and racism.[34] More complex and culture-sensitive (though not unproblematic) analyses had to await the emergence of historical changes in the relations between the genders. These changes have created a massive conflict between the culturally favored forms of producing persons (gendered, raced, classed persons) and the beliefs and actions of increasing numbers of women and some men who do not want to live out mutilated lives within the dangerous and oppressive politics these archaic forms of reproduction encourage.

If we cannot exactly describe this historical moment through an analogy to a "conflict between productive forces and modes of pro-

[30]Harding (1983b). As I shall show, I now have postmodernist questions about my earlier defenses of the standpoint epistemologies.
[31]Engels (1972, 606).
[32]Engels (1972, 624).
[33]O'Brien (1981) also makes this point.
[34]For an analysis of these four main forms of feminism, see Jaggar (1983).

duction" (and why should we have to?), we can nevertheless see clearly many aspects of the specific economic, political, and social shifts that have created this moment. There was the development and widespread distribution of cheap and efficient birth control, undertaken for capitalist and imperialist motives of controlling Third World and domestically colonized populations. There was the decline in the industrial sector combined with growth in the service sectors of the economy, which drew women into wage labor and deteriorated the centrality of industrialized "proletariat" labor. There were the emancipatory hopes created by the civil rights movement and the radicalism of the 1960s in both the United States and Europe. There was the rapid increase in divorce and in families headed by females—brought about in part by capitalism's seduction of men out of the family and into a "swinging singles" lifestyle, where they would consume more goods; in part by women's increased, though still severely limited, ability to survive economically outside of marriage; and no doubt in part by an availability of contraceptives that made what in olden days was called "philandering" less expensive. There was the increasing recognition of the feminization of poverty (probably also an actual increase in women's poverty), which combined with the increase in divorce and the drawing of women into wage labor to make women's life prospects look very different from those of their mothers and grandmothers: now women of every class could—and should—plan for lives after or instead of marriage. There was the escalation in international hostilities, revealing the clear overlap between masculine psychic needs for domination and nationalist domination rhetoric and politics. No doubt other significant social changes could be added to this list of preconditions for the emergence of feminism and its successor science and epistemology.

Thus, to paraphrase Engels, feminist theory is nothing but the reflex in thought of these conflicts in fact, their ideal reflection in the minds first of the class most directly suffering under them—women.[35] Feminist science and epistemology projects are not the products of observation, will power, and intellectual brilliance alone—the faculties that Enlightenment science and epistemology hold responsible for advances in knowledge. They are expressions of ways in which nature and social life can be understood by the new kinds of historical persons created

[35]See Faderman (1981, 178–89) for a valuable analysis of the similar "causes" for the nineteenth-century women's movement in England and America.

by these social changes.[36] Persons whose activities are still characteristically "womanly," yet who also take on what have traditionally been masculine projects in public life, are one such important group of new persons. This "violation" of a traditional (at least, in our recent history) gendered division of labor both provides an epistemically advantaged standpoint for a successor science project and also resists the continuation of the distorting dualities of modernism. Why should we be loath to attribute a certain degree of, if not historical inevitability, at least historical possibility to the kinds of understandings arrived at in feminist science and epistemology?

I still think a historical account is an important component of the feminist standpoint epistemologies: it can identify the shifts in social life that make possible new modes of understanding. A standpoint epistemology without this recognition of the "role of history in science" (Kuhn's phrase) leaves mysterious the preconditions for its own production. However, I now think that the kind of account indicated above retains far too much of its Marxist legacy, and thereby also of Marxism's Enlightenment inheritance. It fails to grasp the historical changes that make possible the feminist postmodernist challenges to the Enlightenment vision as well as to Marxism. We postpone until the next chapter a fuller discussion of this issue.

We saw in Chapter 1 that the feminist empiricist strategy argues that sexism and androcentrism are social biases, prejudices based on false beliefs (caused by superstitions, customs, ignorance, and miseducation) and on hostile attitudes. These prejudices enter research particularly at the stage of the identification and definition of scientific problems, but also in the design of research and in the collection and interpretation of evidence. According to this strategy, such biases can be eliminated by stricter adherence to the existing norms of scientific inquiry. Moreover, movements for social liberation "make it possible for people to see the world in an enlarged perspective because they remove the covers and blinders that obscure knowledge and observation."[37] The women's movement creates the opportunity for such an

---

[36]Chapter 9 outlines the precedents for this kind of analysis in accounts of the breakdown of the medieval division of labor, which permitted the emergence of the new class of craftspeople who created experimental observation in the fifteenth century. See Zilsel (1942) and Van den Daele (1977).

[37]Millman and Kanter (1975, vii).

enlarged perspective and, also creates more women scientists, who are more likely than men to notice androcentric bias.

However, this justificatory strategy undermines key assumptions of its parental empiricist discourse (to paraphrase a point of Zillah Eisenstein's, feminist empiricism has a radical future), and in this undermining—this internal incoherence—we can recognize this epistemology's transitional character and the potential wellsprings of its radicalism.[38]

Feminist empiricism challenges three related and incoherent assumptions of traditional empiricism. First, it questions the assumption that the social identity of the observer is irrelevant to the "goodness" of the results of research, asserting that the androcentrism of science is both highly visible and damaging, and that its most fecund origin is in the selection of scientific problems. It argues that women *as a social group* are more likely than men *as a social group* to select problems for inquiry that do not distort human social experience. Second, feminist empiricism questions the potency of science's methodological and sociological norms to eliminate androcentric biases; the norms themselves appear to be biased insofar as they have been incapable of detecting androcentrism. Third, it challenges the belief that science must be protected from politics. It argues that *some* politics—the politics of movements for emancipatory social change—can increase the objectivity of science. Because the feminist empiricist justificatory strategies reveal the incoherences of traditional empiricism, they also create a misfit, an incoherence, between substantive feminist scientific claims and this feminist epistemological strategy used to justify them.

The recognition of these incoherences led to the development of the feminist standpoint strategies, which appear to be coherent with those elements of feminist empiricism that undermine traditional empiricism. The feminist standpoint epistemologies are grounded in those shared characteristics of *women as a social group* and of *men as a social group* that created feminist empiricism's internal incoherence. But are the standpoint epistemologies internally incoherent along other dimensions?

[38]Eisenstein (1981); she made the point about Liberal feminism. The epistemology which is coherent with Liberal feminism is feminist empiricism.

7 OTHER "OTHERS" AND FRACTURED IDENTITIES: ISSUES FOR EPISTEMOLOGISTS

Now we are in a position to explore dimensions of the internal incoherences in the feminist standpoint epistemological projects. Let us begin by noting that perhaps the proletariat was the only epistemologically advantaged "right group" at the "right place in history" in the nineteenth century. But are women the only such group at this moment in history? If not, what are the intellectual and political relationships between feminist scientific and epistemological projects and the similar projects of the other groups? Furthermore, are women, or even feminists, a "group" in the sense required by the standpoint epistemologies? Do not *other* self-conscious political projects create in many women and feminists self-identities and political loyalties that are in tension with the metaphysics and politics of the standpoint epistemologies?

In short, can there be *a* feminist epistemological standpoint when so many women are embracing "fractured identities" as Black women, Asian women, Native American women, working-class women, lesbian women? Do not these identities undercut the standpoint assumption that common experiences as women create identities capable of providing the grounds for a distinctive epistemology and politics? Even the infamous "hyphenization" of feminist political and theoretical stances—Socialist-Feminism, Radical-Feminism, Lesbian-Feminism, Black-Marxist-Feminism, Black-Lesbian-Socialist-Feminism, Radical–Women-of-Color—bespeaks an exhilaration felt in the differences in women's perceptions of who we are and of the appropriate politics for navigating through our daily social relations. It is an exhilaration similar

163

to the high energies many women initially felt in embracing enthusiastically what had been the degrading label "woman." And it is this "hyphenization" exhilaration that the standpoint epistemologies appear to devalue and make invisible, and that motivates my own ambivalences.[1] The insistence on fractured identities points to the importance of differences in women's politics—whatever our commonalities in experience—which appear to be excluded from the central concerns of the standpoint theories.[2]

At this point, I need to remind the reader that from the theoretical perspective of this study, tensions, contradictions, and ambivalences within and between theories are not always bad. Coherent theories in an obviously incoherent world are either silly and uninteresting or oppressive and problematic, depending upon the degree of hegemony they manage to achieve. Coherent theories in an *apparently* coherent world are even more dangerous, for the world is always more complex than such unfortunately hegemonous theories can grasp. These homilies for a postmodern consciousness are anathema to the modernist consciousness, especially to philosophical modernism; but it is the modernist consciousness that is the problem in this study. The ambivalences within feminism are fruitful guides to the regularities and underlying causal tendencies in the social world within which such theory construction occurs.

My argument is that we should explicitly recognize the ambivalences and contradictions within both feminist and androcentric thinking, and learn how to cherish beneficial tendencies while struggling against the social conditions that make possible regressive tendencies in both. I am not suggesting that we should *try* to produce incoherent theories, but that we should try to fashion conceptual schemes that are more alert to the complex and often beneficial ways in which the modernist world is falling apart.

---

[1] For explorations of the "politics of hyphenization," see Zillah Eisenstein, ed., *Capitalist Patriarchy and the Case for Socialist Feminism* (New York: Monthly Review Press, 1978); and Cherrie Moraga and Gloria Anzaldua, eds., *This Bridge Called My Back: Writings by Radical Women of Color* (New York: Persephone Press, 1981).

[2] Donna Haraway's insistent postmodernist skepticisms over the last few years have helped me to think past my own ambivalences about the standpoint epistemologies. See Haraway (1985) for her feminist postmodernist response to the feminism and science/technology discourses. For a good example of her earlier skepticism, see Haraway (1981). Though we disagree on a number of fundamental issues, I am especially grateful for her critical comments on my treatment of these themes.

A CURIOUS COINCIDENCE

I want to make these issues concrete by exploring the implications of the curious coincidence of African and feminine "world views." I want to begin the task of linking the issues this coincidence raises for the standpoint approaches with those arising in the debates over feminist postmodernism.

We saw earlier that feminists point to a set of conceptual dichotomies within which Enlightenment science and epistemology are constructed: reason vs. emotion and social value, mind vs. body, culture vs. nature, self vs. others, objectivity vs. subjectivity, knowing vs. being. In each dichotomy, the former is to control the latter lest the latter threaten to overwhelm the former, and the threatening "latter" in each case appears to be systematically associated with "the feminine." And while feminist theories have identified a number of aspects of the division of labor by gender that tend to encourage distinctive feminine and masculine world views, one strand in the literature has attributed these distinctive world views to the gendered personalities produced by infants' gender-differentiated experience of the adult division of labor.

Observers of social hierarchies other than that of masculine dominance have pointed to these very same dichotomies as the conceptual scheme that permits these other kinds of subjugation: Russell Means contrasts Native American and Eurocentric attitudes toward nature in these terms; Joseph Needham similarly contrasts Chinese and Western concepts of nature.[3] As we shall see, some observers of both African and Afro-American social life contrast African and European thought in these terms; they posit an African world view which, they imply, could be the origin of a successor science and epistemology. What they call the African world view is suspiciously similar to what in the feminist literature is identified as a distinctively feminine world view. What they label European or Eurocentric shares significant similarity with what feminists label masculine or androcentric. Thus on these separate accounts, people (men?) of African descent and women (Western?) appear to have very similar ontologies, epistemologies, and ethics, and the world views of their respective rulers also appear to be similar.

It is no surprise to be able to infer that Western men hold a distinctively European world view or that the easily detectable expressions

[3] Russell Means, "The Future of the Earth," *Mother Jones* (December, 1980); Needham (1976).

of a European consciousness are masculine. But it is startling to be led to the inference that Africans hold what in the West is characterized as a feminine world view and that, correlatively, women in the West hold what Africans characterize as an African world view.

Furthermore, how should we think of the world view of women of African descent? Is it more intensely infused with the overlapping feminine/African conceptual schemes than is the case for either their African brothers or their Western sisters? This reasonable inference from claims made in the African and feminist literatures flies in the face of repeated observations that Black women, like women in other subjugated racial, class, and cultural groups, have been denied just the degree of femininity insisted upon for women in the dominant races, classes, and cultures. In racist societies, "womanliness" and "manliness," "femininity" and "masculinity," are always racial as well as gender categories.[4] Of course women of African descent, no less than white women, have presumably gone through distinctively feminine processes of development that bear at least some resemblance to the analogous Western processes: the infants' first "caretakers" are primarily women; to become a woman is, at least in part, to become a potential mother, a potential wife, the kind of person devalued relative to men, and so on. The reader can already begin to glimpse the array of conceptual problems generated by looking back and forth between these two virtually discrete literatures.

This chapter considers a number of possible solutions to and ways of dissolving these problems. One likely objection must immediately be addressed. For what appear to be good reasons, white feminists frequently balk at the very idea of a unified world view shared by peoples in the many and very different cultures of Africa and the "African diaspora." Certainly the literature suggesting such a world view—which, though it had colonial origins, is now produced primarily by Africans and Afro-Americans—draws our attention away from important cultural differences. It may even create fictitious com-

---

[4]See Davis (1971); Caulfield (1974); Boch (1983); Hooks (1981; 1983). In racially stratified cultures, androcentrism always prescribes different restrictions for women in the subjugated and dominant races; in gender-stratified cultures, racism takes different forms for men and women. Think of current state pronatalist policies for white middle-class women while supports for child rearing are systematically being withdrawn for poor and black women. Think of the expressions of "femininity" expected of the slaveowner's wife and of his black slave women; of the expressions of "masculinity" expected of the slaveowner and his black slave men; of the different racist restrictions on white men and women and on black men and women.

monalities—but no more so than do feminist accounts that attribute unitary world views to women and men respectively, ignoring differences created by the social contexts of being black or white, rural or industrialized, Western or non-Western, past or present. Moreover, there may well be very general commonalities to be found across all these cultural differences. After all, we are not too uncomfortable speaking of a "medieval world view," a "modern world view," or a "scientific world view," despite the cultural differences in the peoples to whom we attribute these very general conceptual schemes and corresponding ways of organizing social relations.

Before discussing these issues in greater detail, there are two points I wish to make. First, while there are good reasons to be critical of a kind of generalizing that has its roots in colonial projects of imperialism, it would be a problem requiring explanation if there were no significant differences in the world views of the colonizers and the colonized. Second, feminists should be equally critical of overgeneralizing in our own theories.

Here I shall first lay out the other half of the correlation—the African world view—and identify a series of problems with which these world views and their commonalities confront us before exploring some real issues the similar projects of these two literatures raise for both theory and politics at this moment in history.[5]

THE AFRICAN WORLD VIEW

In a paper entitled "World Views and Research Methodology," the Black American economist Vernon Dixon[6] explains why the economic behavior of Afro-Americans is persistently perceived, through the lenses of neoclassical economic theory, as deviant. The "rational economic man" of the European theory, he argues, is in fact only European; aspects of Afro-American economic behavior that appear irrational

[5]See Harding (1986) for my briefer discussion of these issues as they apply to the kinds of claims Gilligan (1982) and others make about the grounding for a theory of women's distinctive moral concerns.

[6]Dixon (1976). See also Hodge, Struckmann, and Trost (1975); Gerald G. Jackson, "The African Genesis of the Black Perspective in Helping," in *Black Psychology*, 2nd ed., ed. R. L. Jones (New York: Harper & Row, 1980), pp. 314–31; and the sources Dixon cites. Subsequent page references to the Dixon paper are in the text; I quote extensively from it to allay suspicions that either Dixon or I use overt gender metaphors to describe the phenomena he examines. The issues raised in these U.S. writings are related in a way too complex to examine here to the discussions of what constitutes African philosophy. See, e.g., Keita (1977–78), Wiredu (1979), Hountondji (1983).

167

from the perspective of neoclassical economic theory appear perfectly rational from the perspective of an African world view.

Dixon locates the major difference between the two world views in the European man-to-object vs. the African man-to-person perception of the relationship "between the 'I' or self (Man) and everything which differs from that 'I' or self.... other men, things, nature, invisible beings, gods, wills, powers, etc., i.e., the phenomenal world." Among Euro-Americans, he says,

> there is a separation between the self and the nonself (phenomenal world). Through this process of separation, the phenomenal world becomes an Object, an "it." By Object, I mean the totality of phenomena conceived as constituting the nonself, that is, all the phenomena that are the antithesis of subject, ego, or self-consciousness. The phenomenal world becomes an entity considered as totally independent of the self. Events or phenomena are treated as external to the self rather than as affected by one's feelings or reflections. Reality becomes that which is set before the mind to be apprehended, whether it be things external in space or conceptions formed by the mind itself. [pp. 54–55]

Dixon cites empirical studies such as one that found in Euro-American students a systematic "perception of conceptual distance between the observer and the observed; an objective attitude, a belief that everything takes place 'out there in the stimulus.' This distance is sufficiently great so that the observer can study and manipulate the observed without being affected by it" (p. 55).

The fundamental Euro-American separation of the self from nature and other people results in the objectifying of both. The presence of empty perceptual space surrounding the self and separating it from everything else extracts the self from its natural and social surroundings and locates all the forces in the universe concerned to further the self's interests inside the circle of empty perceptual space—that is, in the self itself. Outside the self are only objects that can be acted upon or measured—i.e., known. Nature is an "external, impersonal system" which, since it "does not have his interest at heart, man should and can subordinate . . . to his own goals." "The individual becomes the center of social space," and so "there is no conception of the group as a whole except as a collection of individuals." Thus "the responsibility of the individual to the total society and his place in it are defined in terms of goals and roles which are structured as autonomous." "One's rise up the ladder of success is limited only by one's individual talents. Individual effort determines one's position" (p. 58).

This conception of the self as fundamentally individualistic also limits one's obligations and responsibilities. "One retains the right to refuse to act in any capacity. It is not expected that a man, in pursuing his own goals of money-making and prestige, will remain dedicated to the goals of a given firm, college, or government agency if he receives an offer from another institution which will increase his salary or status. The individual only participates *in* a group; he does not feel *of* the group. In decision-making, therefore, voting rather than unanimous consensus prevails" (pp. 58–59).

In the Africanized world view, there is no gap between the self and the phenomenal world: "One is simply an extension of the other." For people with this kind of ontology, there is

a narrowing of perceived conceptual distance between the observer and the observed. The observed is perceived to be placed so close to the individual that it obscures what lies beyond it, and so that the observer cannot escape responding to it. The individual also appears to view the "field" as itself responding to him; i.e., although it may be completely objective and inanimate to others, because it demands response it is accorded a kind of life of its own. [p. 61]

Given this conception of the self and its relationship to the phenomenal world, Africans

experience man in harmony with nature. Their aim is to maintain balance or harmony among the various aspects of the universe. Disequilibrium may result in troubles such as human illness, drought, or social disruption. . . . According to this orientation, magic, voodoo, mysticism are not efforts to overcome a separation of man and nature, but rather the use of forces in nature to restore a more harmonious relationship between man and the universe. The universe is not static, inanimate or "dead"; it is a dynamic, animate, living and powerful universe. [pp. 62–63]

Furthermore, "the individual's position in social space is relative to others. . . . The individual is not a human being except as he is part of a social order." "Whatever happens to the individual happens to the whole group, and whatever happens to the whole group happens to the individual." In this communal rather than individualistic orientation, "an individual cannot refuse to act in any critical capacity when

169

called upon to do so." Thus Afro-Americans will often "unquestioningly go against their own personal welfare for other Blacks . . . even though the former know that the latter are wrong. They will co-sign loans for friends while aware that their friends will default and that their own finances will suffer." An orientation toward interpersonal relationships has predominance over an orientation to the welfare of the self (pp. 63–64).

In knowledge-seeking, the European first separates himself from what is to be known, then categorizes and measures it in an impartial and dispassionate manner. Africans "know reality predominantly through the interaction of affect and symbolic imagery," which—in contrast to intuition—requires inference from or reasoning about evidence. But in contrast to European modes of gaining knowledge, it refuses to regard what is known as value-free, or to see either the knower or the process of coming to know as impartial and dispassionate. Feelings, emotions, and values are regarded as a necessary and positive part of coming to know (pp. 69–70).

In summary, Dixon argues that the African world view is grounded in a conception of the self as intrinsically connected with, as a part of, both the community and nature. The community is not a collection of fundamentally isolated individuals but ontologically primary. The individual gets his sense of self and can determine what it is only through his relationships within a community. His personal welfare fundamentally depends upon the welfare of the community, rather than the community's welfare being dependent upon and measurable in terms of the welfare of the individuals who constitute it. Because the self is continuous with nature rather than set over and against it, the need to dominate nature as an impersonal object is replaced by the need to cooperate in nature's own projects. Coming to know is a process involving concrete interactions that acknowledge the role of emotions and values in gaining knowledge, and recognize the world-to-be-known as having its own values and projects.

COMMONALITIES AND PROBLEMS

There are differences between the African vs. European and feminine vs. masculine dichotomies—not so much between the world views attributed to Europeans and men as between those attributed to Africans and women. This should not be surprising since there are important differences between the life worlds of Africans and Afro-

Americans on the one hand, and women of European descent on the other (I return to this point later). Nevertheless, the similarities are striking.[7]

Europeans and men are thought to conceptualize the self as autonomous, individualistic, self-interested, fundamentally isolated from other people and from nature, and threatened by these others unless the others are dominated by the self. Both groups perceive the community as a collection of similarly autonomous, isolated, self-interested individuals having no intrinsically fundamental relations with one another. For both groups, nature is also an autonomous system from which the self is fundamentally separated and which must be dominated to alleviate the threat of the self's being controlled by it.

To Africans and women are attributed a concept of the self as dependent on others, as defined through relationships to others, as perceiving self-interest to lie in the welfare of the relational complex. Communities are relational complexes that are ontologically and morally more fundamental than the persons that are individuated through their positions in the community. Nature and culture are inseparable, continuous.

From these contrasting ontologies follow contrasting ethics and epistemologies. To Europeans and men are attributed ethics that emphasize rule-governed adjudication of competing rights between self-interested, autonomous others; and epistemologies that conceptualize the knower as fundamentally separated from the known, and the known as an autonomous "object" that can be controlled through dispassionate, impersonal, "hand and brain" manipulations and measures. To Africans and women are attributed ethics that emphasize responsibilities to increasing the welfare of social complexes through contextual, inductive, and tentative decision processes; and epistemologies that conceptualize the knower as a part of the known, the known as affected by the process of coming to know, and that process as one which unites hand, brain, and heart.

Feminists and Africanists are clearly onto something important.

[7]See Gilligan (1982) for the classic discussion of women's different moral voice. See Kittay and Meyers (1986) for discussions by philosophers of the issues Gilligan raises. The feminist object-relations theory, discussed in Chapter 5, perhaps provides the most explicit account of gendered ontologies, epistemologies, and world views. See Chodorow (1978), Dinnerstein (1976) and Flax (1983) for development of the theory; see Balbus (1982), Ruddick (1980), Harding (1980; 1981; 1982), and many of the essays in Harding and Hintikka (1983) for samples of the widespread uses of this theory by feminists.

171

However, there are many problems with taking the claims in the two literatures at face value; indeed, recognition of the similarities intensifies the already severe conceptual problems within each literature, some of which are analogous.

Residues of Colonial and Patriarchal Conceptual Schemes.

As we noted at the beginning of this chapter, some Westerners, conscious of the way Euro-American imperialism shaped African social life, are loath to countenance generalizations about an African character or world view. Some people of African descent make the same criticism. For these thinkers, such generalizations smack of the politics and mystifications of racism.

They point out that the term "African" does not pick out the way the peoples who live on that continent have named themselves, nor the way that they primarily identify themselves even today. As the Afro-American philosopher Lancinay Keita explains, the concept of "Africa" first appeared in European writings of the early modern period.[8] It was a way to simplify thinking about a group of peoples who were about to be "other"—that is, exploitable with impunity—as far as the imperialist projects of Europeans were concerned. The ways humans divide up the surface of the globe are political; there is nothing "natural" about how we conceptualize the boundaries of states or continents. The political origins of the concept of Africa as a geographical unit are clearly traceable to the period when Europeans and people of European descent began removing from Africa pieces of the physical land and the congealed labor of Africans in the form of diamonds, raw materials, and commodities, as well as the peoples themselves.

Consequently, using the concept today for analyses whose goal is the emancipation of the ex-colonized is problematic; its very use paradoxically reinforces the legitimacy of the European tradition of insisting on the right to name, and therefore to treat, non-Europeans in ways that serve European interests. In particular, it emphasizes the "otherness," the alienness, of the ontologies, epistemologies, and ethics of people of African descent relative to those of people of European descent. It reinforces the contrast paradigm that has been so useful in projects of domination.

A correlative problem occurs in the feminist literature. "Woman" and "femininity" are concepts created through and central to masculine

[8]Keita (1977–78).

172

dominance projects. Once we recognize that gender differences are socially created, we notice that only within the cultural projects of masculine-dominated societies does it become important to emphasize what we can now see are cultural differences between the genders and to insist on the fundamental sameness of women in every culture. Masculinist conceptual schemes lead us to think that men are endlessly fascinating in their individual and collective historical particularity— this is what makes the "men's history" we have appear to men to be human history—but that women in every race, class, and culture are always best understood as members of a "sex-class," of a gender class. Women are not historically—that is, socially—interesting at all; women's contributions to human history are entirely defined by their role in reproduction. Thus to focus on *women's* world view, or the *feminine* world view, paradoxically supports a masculinist conceptual scheme.

Ahistoricity.
The first problem is related to a second. The concept "African" tends to paper over the vast differences between the histories and present projects of the hundreds of indigenous African cultures. Thinking only of the presumed commonalities between the peoples of West Africa and East Africa, or between persons of color from the Caribbean and Chicago, creates a reality that may be largely fictional. "Bantu ontology" is, perhaps, one thing; "African ontology" is quite another.[9]

Moreover, the concept suggests that the presence of European rule on the African continent has left unchanged the presumably ancient ways of understanding self, others, and nature. Historical studies reveal that this is most certainly not true. The ontologies, theories of knowledge, and ethics of a culture always change over time. And the experience of colonization, whether for people in Africa or for those forcibly removed to the colonizing cultures, would exacerbate such change. Furthermore, it is not just "normal" historical change that is problematic here. The reports upon which claims of an African world view are primarily based derive from studies of cultures that have been

---

[9]Placide Tempels, *Bantu Philosophy* (Paris: Presence Africaine, 1959). But see Hountondji (1983) and Wiredu (1979) for criticism of this kind of account and its claims to constitute "African philosophy." Abiola Irele's introduction to Hountondji's essays provides a useful review of the history of these African ethnophilosophies. And the notion of "ethnophilosophies" is useful for thinking about feminist explorations of the "indigenous" feminine world view. Is it also useful for thinking about the hegemonous status in the West of the world view developed by Descartes, Hume, Locke, and Kant? Wiredu (1979) thinks so, and so do the feminist science critics.

struggling against and within Western imperialism, Western conceptual schemes and social structures, for centuries. How much of the reported African world view is "indigenous" and how much has been shaped by such struggles?

Similarly, to focus on the women's world view obscures and mystifies the vast differences between women's experiences and characteristics in different cultural settings. How significant can the commonalities be between the concepts of self, other, nature, and their interrelations held by medieval peasant women and either nineteenth-century feminist or antifeminist women; by women members of the "industrial proletariat" and women professionals; by Black women in New York City and women in still existing gathering societies?

Moreover, the concept "women" suggests a fundamental continuity between, before, and during the historically specific (even if incredibly long) history of men's subjugation of women. But anthropologists point to the generally lower "quality of gender" in cultures with lower division of labor of any other kind. Some insist that it was only the contact with and eventual incorporation into Western capitalism that substituted masculine-dominant for "egalitarian" societies.[10] Another argues that gender itself emerged, complete with its asymmetry of masculine dominance, only within distinctively human history.[11] It is clear that these accounts conflict with each other; furthermore, "the dawn of human history" is capable of providing only fragile evidence for any claims, as noted earlier. Nevertheless, the anthropological accounts tend to understand gender as the "relational and contextual construct" we saw Jane Flax call for.[12] How much of women's world view is common to women through all these changes in the relations and contexts of the history of gender? How much is the product of struggle against masculine dominance and how much of women's different embodiment and early gendering processes?

Contrast Schemas.
Feminine vs. masculine and African vs. European are contrast schemas,[13] originating primarily in men's and Europeans' attempts to define

[10]E.g., Leacock (1982).
[11]Cucchiari (1981).
[12]Flax (1983; 1986).
[13]Horton (1973) and Hountondji (1983) provide good criticisms of the problems with such schemas. Wiredu (1979) points to some popular contrasts Horton (1967) failed to question in an otherwise pathbreaking analysis of the commonalities between Western science and African traditional thought.

174

as "other" and subhuman the groups they chose to subjugate.[14] The original social process of creating the genders is lost to our view in the distant mists of human history, but the original creation of races is entirely visible in relatively recent history.[15] And apart from origins issues, we can see both kinds of difference under constant reconstruction in our past and present. Differences between the races in the United States have been brought about through political processes of slavery; Native American genocide; nineteenth- and early twentieth-century immigration, labor, and reproductive policies; and continuing institutionalized anti-Semitism and other forms of racism. Historians describe similar political processes that have simultaneously legitimated and created modern forms of observable gender differences.

There are four points to be made here. First, racial and gender contrast schemas originate within projects of social domination. Therefore, we should look to the history of those projects to locate the primary causes of subsequent differences between the races and genders. I suggest that when we do, we will notice that it is the same group of white, European, bourgeois men who have legitimated and brought into being for the rest of us life worlds different from theirs. In this sense, it is one contrast schema we have before us, not two. And it is not one primarily of our making, either ideologically or in lived experience.

Second, any contrast schema over-emphasizes certain differences at the expense of others and of commonalities that are both just as "real." Is it observable differences between men and women we want to emphasize in feminist theory, or rather differences between the social projects and fantasy lands of bourgeois, Western, antifeminist men and the projects and hopes of the rest of us?

Third, such schemas also exalt intragroup commonalities at the expense of differences. The masculine and Eurocentric world view appears more coherent than do the collective world views of those it

---

[14]An interesting and important question is, to what extent Africans and women also participated in the conscious or unconscious construction of these contrast schemas— but as acts of rebellion against the hegemony and value accorded to the European and masculine, respectively. With reference to the gender contrast, I refer not to such processes as women's participation in hoisting themselves onto the proverbial pedestals but to the little daily appreciations of "the feminine" as both a strength and a refuge in a heartless world, which are common among mothers and daughters, sisters, and women friends.

[15]See Cucchiari (1981) and Rubin (1975) for contrasting attempts to probe those mists, and Keita (1977–78) for the origins of the African vs. European contrast.

defines as "other." We are assigned to different subjugations by our single set of rulers, and these differences occur within women's history and African history as well as between the two.

Fourth, while there is no denying that men and women in our culture do live in different life worlds, there is something at least faintly anachronistic about our emphasis on these differences during a period when they are presumably disappearing for many of us. Imagine how much greater these differences were in the gender-segregated lives of the nineteenth-century bourgeoisie. As the divisions of human activity and social experience that created men and women (in the nineteenth-century, bourgeois sense of these terms) disappear, should we not expect feminine and masculine world views in these groups to begin to merge? Should we not ask similar questions about the African vs. European contrast?

These contrast schemas are valuable for identifying the far less than emancipatory aspects of the Western world view within which we are all supposed to want to live out our lives (and, inadvertently, for identifying what to many of us are undesirable aspects of "the feminine" or "the African"). Focusing on women's and Africans' different realities clarifies how far from emancipatory that world view is. My cautions here are about tendencies to exalt women's different reality when it is also less than the reality we want, is not the only alternative reality, and is disappearing.

Metaphoric Explanation.

Race and gender metaphors have often been used to explain other phenomena. The behavior of Africans, Afro-Americans, Native Americans, and other racially dominated groups; male homosexual behavior; and the reproductive behavior of females (and sometimes even of males) among the apes, sheep, bees, and other subhuman species have all been characterized as "femininized."[16]

That is not happening in either of the literatures we are considering. But a more subtle kind of metaphoric explanation may be occurring: namely, differences that correlate with gender difference are conceptualized as gender differences; those that correlate with race difference are conceptualized as racial differences. For instance, because women in our culture tend to have an ethic of caring rather than of rights,

---

[16]For criticisms of this practice, see, e.g., the papers in Hubbard, Henifin, and Fried (1982).

this is conceptualized as feminine. As we all know, correlation is not the most reliable form of explanation. If men of African descent also tend toward an ethic of caring rather than of rights, we need to look beyond Western women's distinctive social experiences to identify the social conditions tending to produce such an ethic.

Our own gender totemism obscures for us the origins of the gender dichotomies we observe. What is interesting about the totemism that anthropologists describe is the relationship not between the signifier and the thing signified but between the signifiers. It is not the two relationships between one tribe and wolves and another tribe and snakes that anthropologists have found revealing but the relationship between the meanings of wolves vs. snakes for both tribes.[17] Similarly, attention to gender totemism in the attribution of gendered world views leads us to examine the meanings of masculinity and femininity for men and women rather than the fit between these meanings and observable beliefs and behaviors. The contrast schemas exacerbate the tendency within feminism to preoccupation with gender symbolism at the expense of the complex realities of only "hyphenatedly" gendered social structures and individual identities.

Each Literature Denies the Other's Dichotomy.
The preceding four problems conjoin to create a fifth, which is the most important one for motivating interest in a feminist postmodernist consciousness. Where the Africanists find important differences between the world views of peoples of African and European descent, and feminists find important differences between the world views of women and men within Western cultures, neither acknowledges the other's dichotomy within its own conceptual scheme. Thus we are encouraged to assume that there are no significant differences between the concepts of self, community, and nature for men and women of African descent; no significant differences in these concepts between women of African and of European descent in America today. These assumptions are damaging to the adequacy, not to mention the political appeal, of each analysis.[18]

[17]Judith Shapiro pointed this out in "Gender Totemism and Feminist Thought," a paper presented at the University of Pennsylvania Mid-Atlantic Seminar for the Study of Women and Society (October 1984). Shapiro's questions at the November 1983 meeting of the seminar brought home to me the importance of sorting out the conceptual problems in the discourses under consideration here.

[18]While trying to avoid moralisms, I must remind the reader that my own work over the past few years is a target of my ambivalences; see, e.g., Harding (1980; 1981; 1982;

177

However, this problem originates in large part in the strengths, in the importance, of the history each theory focuses on. From the perspective of the feminist literature, in African and other preindustrialized, "traditional" societies, the gender order has been a fundamental component of the social order. Consequently, it is African men who have been the leading spokespersons for African liberation movements. In the United States, Black men have taken a disproportionate number of the leadership roles within Black communities, and if a white social structure that is sexist as well as racist is going to listen to Black Americans at all, it is Black men they will be willing to hear. From the perspective of the African literature, feminism has been constructed within a social order that is racist as well as sexist. White women have taken a disproportionate number of the leadership roles within feminist politics. It is white women whose perspective on women's lives is primarily sought when the topic is social relations between the genders. Black women are beginning to be encouraged to speak about their own lives, but only white women about "women's lives."

As a consequence of these excessive universalizations in the analyses of Africanists and feminists, women of African descent, in the United States and in Africa, totally disappear from both analyses, conceptualized out of existence because African men and white women are taken as the paradigms of the two groups. Their critical perceptions and political leadership are delegitimated because they are of African descent, and as African because they are women. This is especially ironic since women constitute at least half of the people of African descent, and women of color constitute the vast majority of women in the world. For neither literature is it even a "minority" which is excluded. To the extent that each liberation project implies that women of African descent do not exist, we fail in our own projects; we think and struggle within conceptual schemes that prevent the majority of our purported constituencies from benefitting from the goals of our struggles. More accurately, the goals themselves are regressive insofar as we exclude women of color from a central role in defining them. As the editors of one collection about Black women's studies put it, "All the women are white, all the Blacks are men, but some of us are

---

1983b), and what I still think (insist!) are the brilliant and provocative essays in Harding and Hintikka (1983). As I shall explain further, I am not willing to give up entirely what I think are important theoretical and political advantages of the standpoint epistemologies. My approach differs in this respect, I think, from Haraway (1981; 1985).

brave."[19] This problem alerts us to procede cautiously in drawing inferences about why a "women's world view" and an "African world view" appear to coincide.

IMPROBABLE EXPLANATIONS

Before considering some possibly fruitful explanations for the curious coincidence of African and feminine world views, we need to examine three implausible ones that have already surfaced in our discussion of these literatures.

Not Biology.
First of all, it should be obvious that appeal to biological differences cannot account for the similar dichotomies. The problem here is the biological determinist tendencies not only within European and sexist thinking but also within Africanist and feminist thinking. Thinking with "one ear tuned to each literature," so to speak, reveals how poor we all are at conceptualizing the relationship between biology and culture. We are indeed embodied social persons, not disembodied asocial minds as the Cartesian tradition holds. So our differently embodied interactions with our surroundings should create different experiences and thus different kinds of beliefs. But how these biological differences constrain experience and thus belief is apparently something we do not yet know how to conceptualize.

Some writers of African descent claim that the higher presence of melanin in Black peoples provides the physiological basis for "psychological oneness"; others propose that differences between Caucasians and Blacks in the patterns of the amino acids to be found in urine, in the consistency of ear wax, and in brain patterns underlie cultural differences between Africans and Europeans.[20]

Similarly, some feminist theorists have argued that physiological differences between females and males ground gender differences. Females *should* have closer relations to others and to nature than do males, and thus culture simply elaborates biological difference. The lines where self ends and "other" begins, where the cultural ends and the

[19]Hull, Scott, and Smith (1982).
[20]Dubois Phillip McGee, "Psychology: Melanin, The Physiological Basis for Psychological Oneness," in L. M. King, V. J. Dixon, W. W. Nobles, eds., *African Philosophy: Assumptions and Paradigms for Research on Black Persons* (Los Angeles: Fanon Center, Charles R. Drew Postgraduate Medical School, 1976).

"natural" begins, are less defined for women because of the nature of female bodies. Furthermore, women's bodies seem to violate deep and universal cultural taboos. In menstruation, women bleed but do not die; in intercourse, women's bodily boundaries are crossed, "violated," though pleasure results; in pregnancy, another human lives inside a woman's body; in nursing, another human eats from a woman's body. These taboos have been constructed in ways that take men's as the ideal human bodies; nevertheless, they mark important differences between male and female bodies that produce different kinds of social experiences and therefore, it is claimed, provide bases in embodiment for sex-differing beliefs. Thus it is not surprising that culture can construct relational vs. separation-maintaining personalities on the foundation of biological sex difference; even if the cultural constraints on gender ceased to exist, men and women would still see the world in residually different ways because of their biological differences.[21]

Whatever plausibility these arguments have with respect to the cognitive dichotomy each one is intended to explain appears to be deteriorated by the similarities between the two dichotomized world views. Since Africans differ by sex and women by race (among other ways), the coincidence of African and women's world views suggests that biology plays a virtually negligent role in the construction of these dichotomies. "Added" together, the grounding of these two kinds of differences in biology could make sense at most for European men and women of African origin. For women of European origin or men of African origin, the two sets of assumed biological groundings would conflict.

Furthermore, to insist on the ability of contemporary science—and

[21]Followers of Lacan appear to be arguing against their object-relations colleagues that no variety of "alternative parenting"—by fathers, homosexual coparents, single mothers—can overcome the effects of the mother-child bond or the father's phallic presence. However, even the object-relations theorists sometimes intimate that biology is to blame for "sexual arrangements and human malaise." Dinnerstein (1976), e.g., discusses the legacy the "obstetrical dilemma" at the dawn of human history left for human gender relations. Mary O'Brien (1981) insists that biology is not destiny, but her account of differences in our consciousness of our reproductive systems suggests a biological basis for gender ideology. Whether writing within the assumptions of feminist theory or not, Jean Elshtain and Carol MacMillan think feminism would be advanced by a better understanding of the significance of biological difference. See Jean Elshtain, "Feminists against the Family," *Nation*, Nov. 17, 1979; *Public Man, Private Woman: Women in Social and Political Thought* (Princeton, N.J.: Princeton University Press, 1981); "Antigone's Daughters," *Democracy* 2(no. 2) (1982); and Carol MacMillan, *Woman, Reason and Nature* (Princeton, N.J.: Princeton University Press, 1982).

perhaps any future science—to identify the distinctively "natural" components of the traits and behaviors of humans is to tread on very thin empirical and theoretical ground. After all, every trait and behavior we can observe in the humans around us is inextricably shaped by culture. From conception to death, differences in our bodies are formed not only by genetic inheritance but also by the food we eat, the air we breathe, the kind of work we do, and other social practices. Even our genetic inheritances are in part the product of social factors, since the "mating" essential to creating a genetic inheritance is itself shaped by social factors. With respect to race differences, it has proved impossible to identify the purely biological components in any trait or behavior.[22] The best we can do is to identify elements of traits and behaviors that are more or less susceptible to social manipulation. The less susceptible we call "biological" or "natural"; the more susceptible, "cultural." Yet even the modern scientific form of this distinction is a cultural creation. It does not unambiguously match the way other cultures draw the distinction.[23]

We are on firmer ground when it comes to *perceptions* of biological differences, for it is plausible to assume that these are entirely culturally determined. The division of humans into races is a cultural act, and how the division is made is extremely variable historically. Similarly, the division of humans into two or more sexes depends upon a culture's interest in and ability to perceive sex differences at all, as well as upon what they are taken to consist in. Recently, these perceptions have changed as genetic, hormonal, and other physiological criteria have joined "gross morphological differences." The sex research literature is full of cases where the various indicators of sex difference do not neatly line up to produce individuals who are—by whatever current criteria—unambiguously male or female, though the individuals often experience themselves as unambiguously sexed. Thus the perception that nature created only two sexes, or that two is the natural number and more than two the result of biological mistakes, is culturally shaped.

[22]Jean Hiernaux, "The Concept of Race and the Taxonomy of Mankind," in *The Origin and Evolution of Man: Readings in Physical Anthropology*, ed. Ashley Montagu (New York: Crowell, 1964), pp. 486–95; Frank Livingstone, "On the Nonexistence of Human Races," *Current Anthropology* 3 (1962): 279–81. See also the discussion of this issue in Cucchiari (1981).

[23]See the essays in MacCormack and Strathern (1980), especially Maurice and Jean Bloch's "Women and the Dialectics of Nature in Eighteenth Century French Thought," which shows how the dichotomy had no fixed referents even in this local domain of Western thought.

Finally, at least one anthropologist argues that the perception of any kind of sex difference between humans is most likely a cultural emergent. It may appear improbable to us that our human ancestors did not conceptualize the pattern of "gross morphological difference" by sex as responsible for differing male and female participation in childbearing, but perhaps they had no reason to do so. After all, initially indiscriminate human desire for bodily contact (Freud's "polymorphous perversity" of infantile sexuality) combined with continuous "sexual receptivity" in females could have permitted the species to reproduce without conceptualizing sex difference.[24] Perhaps our distant descendants will conceptualize the three-millennia-or-so period of human history in which we live as the "Era of Obsession with Sex Difference!"

These considerations reveal that we are a long way from an adequate conceptualization of the constraints that biological differences set on belief patterns. In particular, there appears to be a gap between the understanding that race differences and sex differences are themselves cultural constructs, and the differing conceptions of self, others, and nature that are claimed to divide races and genders. The problems in each literature are exacerbated by the similarities between the two dichotomized sets of beliefs.

Not "Folk Thought" vs. Scientific Thought.
Westerners have frequently described the contrast between African thinking and Post-Enlightenment European thinking as a contrast between "folk thought" and scientific thought, superstitious and critical thought, prelogical and logical thought.[25] The similarities between African and women's world views might tempt one to conclude that such dichotomies explain the correlation: the feminine world view, too, expresses folk, superstitious, or prelogical thinking.

However, this temptation should be resisted. In the first place, the definition of logical and rational thinking is itself a cultural artifact that has changed even within the history of Western thought. Judgments of logicality and rationality are "theory-dependent": what counts as a logical statement depends upon other views a society holds about self, community, nature, and their relationships. The beliefs that appear

[24]Cucchiari (1981). See also Caulfield (1985).
[25]The writings of the French anthropologist Lucien Lévy-Bruhl provide the paradigm case here. See the discussion of this issue in Horton (1967; 1973); Hountondji (1983); Wiredu (1979).

logical to one who conceptualizes species as related to each other through evolutionary patterns will differ from those of one who conceptualizes species as all created by God in the first week of the universe. Different claims will be judged logical by one who conceptualizes the motions of a planet as caused primarily by the gravitational pull of the sun than by one who does not make such assumptions. One project of Thomas Kuhn's *Structure of Scientific Revolutions* is to show the rational, logical processes of reasoning which (together with the best empirical observation possible at the time) led to subsequently discredited scientific claims.[26] Criteria for logicality have shifted within Western science—even in mathematics, as we saw in Chapter 2. As Quine argued, we should give up the dogmatic belief that there is a fundamental cleavage between the truths of logic and those of science. Furthermore, anthropologists have demonstrated the logical, rational character of belief patterns that appear bizarre to Westerners. Robin Horton shows that African traditional thought uses the very same kinds of explanatory strategies as Western science to make sense of the world.[27] So it can not be logical or rational thinking that distinguishes European from African or masculine from feminine world views.

Nor is it critical thinking vs. folk thought—at least not in the way the difference has been understood by Westerners. Ghanian philosopher Kwasi Wiredu agrees with anthropologists like Horton that the logical/prelogical, rational/irrational distinction by which Westerners attempt to account for the differences between African and European thinking is actually a consequence of the fact that many Western anthropologists have been unfamiliar with the patterns of theoretical thinking in Western science. But Wiredu argues that Horton, like other Westerners, overlooks the ubiquitous presence of folk thought in Western cultures: if we take as the mark of critical thought an adherence to "the principle that one is not entitled to accept a proposition as true in the absence of any evidential support," then critical thinking

is not Western in any but an episodic sense. The Western world happens to be the place where, as of now, this principle has received its most sustained and successful application in certain spheres of thought, notably in the natural and mathematical sciences. But even in the Western world there are some important areas of belief wherein the principle does not hold sway. In the West just as anywhere else the realms of religion,

[26]Kuhn (1970).
[27]Horton (1967).

183

morals and politics remain strongholds of irrationality. It is not uncommon, for example, to see a Western scientist, fully apprised of the universal reign of law in natural phenomena, praying to God, a spirit, to grant rain and a good harvest and other things besides.[28]

In Chapter 2, we saw that scientific thinking itself incorporates what are probably best regarded as mystical elements. Thus Western thinking, even as exemplified by men of science, is often "irrational" and uncritical "folk thought." On the other hand,

> no society could survive for any length of time without conducting a large part of their daily activities by the principle of belief according to the evidence. You cannot farm without some rationally based knowledge of soils and seeds and of meteorology; and no society can achieve any reasonable degree of harmony in human relations without a *basic* tendency to assess claims and allegations by the method of objective investigation. The truth, then, is that rational knowledge is not the preserve of the modern West nor is superstition a peculiarity of African peoples.[29]

Similarly, one major trend in feminist analysis has been to show that women's distinctive social experience provides them with evidence for beliefs that have appeared irrational and uncritical to men. Feminist writings have pointed again and again to the irrational, illogical, uncritical "folk" elements in masculine thinking, and to the rational, logical, critical elements in distinctively feminine thinking. Certain emotions and feelings are good reasons for beliefs and actions. "Maternal thinking" draws on different evidence than does paternal thinking. Men's conceptual schemes and problematics simply "fit" the conceptual schemes and problematics suitable for administrative forms of ruling.[30]

Finally, Wiredu points out that for many Western philosophers, reasoning grounded in the assumptions of British empiricism or any other *traditional* Western philosophical framework is now closer to folk thought than to rational, critical thinking.[31] We can add that this evaluation holds, too, for reasoning based on the vast majority of assumptions that make up "the scientific world view." Most of the beliefs of the average or even extraordinary Western scientist or intellectual are

[28]Wiredu (1979, 136).
[29]Wiredu (1979, 137).
[30]See Gilligan (1982); Ruddick (1980); Smith (1974; 1977; 1979; 1981).
[31]Wiredu (1979, 145).

grounded in the "authority of the ancients" rather than in critical, individual evidence gathering. And the (dare one say) fanaticism with which challenges to these beliefs are resisted supports the premise that the fundamental assumptions of the scientific world view are held on the basis of a faith that functions to define the believers' location in a moral/political universe rather than on the basis of critical thinking.

If these dichotomies do not hold up as accurate ways of contrasting European with African world views, then we will want to avoid using them to explain either the contrast between masculine and feminine world views or the overlap between African and women's world views.

Not Entirely a Consequence of Gender Relations.
Both radical feminists and feminist object-relations theorists often state or imply that gender domination is the fundamental human domination, that it produces other forms by providing a first psychological model. They point out that the first human division of labor (beyond, obviously, age division) was by gender, thus paving the way for an asymmetrical gender system among our most distant human ancestors. However, even if gender domination did serve as the original model for other forms of domination, there were many forms by the time humanity reached the period of Euro-American imperialism. In particular, class domination would seem a more likely model for the division of labor between the imperialist nations and the African peoples.

And we certainly cannot explain the African vs. European dichotomy by appeal to the infant's experience of the division of labor by gender. According to the African literature, African men do not have the world view characteristic of Western men, yet African men, too, presumably go through separation and individuation crises which—as long as their early caretakers are primarily women—do not significantly differ from the forms these crises take for Western men. Thus it cannot be the division of labor by gender alone that creates "objectifying" vs. "relational" world views.[32]

To summarize the argument so far, there are striking similarities in the world views attributed to women vs. men and people of African descent vs. those of European descent. But the temptation to hasty generalization about the nature and causes of these similarities must be checked by careful consideration of the conceptual problems in the

[32]But see the attempt to historicize infantile separation and individuation crises, and a suggestion for how to retrodict the emergence of historically varying attitudes toward nature and authority in Balbus (1982, ch. 9).

literatures within which these claims are made, by recognition of the failure of each perspective to acknowledge the existence of the other dichotomy, and by resistance to illicit and undesirable inferences to which one might be led by existing explanatory tendencies.

TOWARD A UNIFIED FIELD THEORY

Thinking about this curious coincidence directs us to seek explanations of observable gender differences different from those we have favored. What we need is something akin to a "unified field theory": a theory that can account for both gender differences and dichotomized Africanist/Eurocentric world views. Such a theory will certainly be an intellectual structure quite as impressive as that of Newton's mechanics, for it will be able to chart the "laws of tendency of patriarchy," the "laws of tendency of racism," and their independent and conjoined consequences for social life and social thought. I make no pretense to the ability to formulate such a theoretically and politically useful conceptual apparatus, but I can point to three analytical notions that illuminate different causal aspects of the correlated dichotomies, and out of which might be constructed the framework for a comprehensive social theory.

Categories of Challenge.
Historians have suggested that "the feminine" functioned as a "category of challenge" in eighteenth-century French thought. We might think of both "the feminine" and "the African" as "categories of challenge."[33] They were in the first place but mirror images of the culturally created categories "men" and "European." They had no substantive referents independent of the self-images of men and Europeans: women were "not-men"—they were what men reject in themselves; Africans were "not-European"—they were what Europeans rejected in their own lives. (Perhaps these categories also express what women and Africans, respectively, claimed for themselves as unappropriatable by the increasing hegemony of a masculinized and Eurocentric world view.) As categories of challenge, the feminine and African world views name what is absent in the thinking and social activities of men and Europeans, what is relegated to "others" to think, feel, and do; what makes possible genderized and racial social orders. In the calls of both for

---

[33]Bloch and Bloch (1980).

sciences and epistemologies, ethics and politics that are not loyal to gender or race dominance projects, we can see in Africanism and feminism "the return of the repressed."

While this notion illuminates ideological aspects of the world views characteristic of Western men and the various groups making up "the rest," it needs to be supplemented by more concrete accounts of the differences in social activity and experience that make the dichotomized views appropriate for different peoples. The other two notions are useful for this task.

Conceptualizers vs. Executors.
Marxists point out that it is the separation of the conception and execution of labor within capitalist economic production that permits the bourgeoisie to gain control of workers' labor.[34] Craft laborers know how to make a pair of shoes or a loaf of bread, but in industrialized economies this knowledge of the labor process is transferred to the bosses and the machines. Capitalist industrialization has increasingly suffused all human labor processes, so that now not only the things made in factories but also such products as the results of scientific inquiry, social services, the enculturation of children, gender relations, and even the meals produced within the household are produced by industrialized processes (see Chapter 3).

Awareness of the increasing division of labor between conceptualizers and executors illuminates the shared aspects of African and women's labor. Imperialism can be understood as enforcing the transfer to Europeans and Americans of the conceptualization and control of the daily labor of Africans. The construction of an ideology that attributed different natures and world views to Europeans and Africans occurred as an attempt by Europeans and Americans to justify this imperialism; the ideology "justified" the exploitation. Prior to the European arrival in Africa, vast trade networks had been organized by Africans; influential centers of African Islamic scholarship existed— Africans had conceptualized and administered a variety of pan-African activities. With the coming of imperialism to Africa, decisions about what labor Africans would perform and who would benefit from it were wrested from Africans and transferred to Europeans and Americans. Henceforth, Africans would work to benefit Euro-American societies, whether as diamond miners, as domestic servants, as the

[34]Braverman (1974).

187

most menial of industrial wage laborers, or as wage or slave labor on plantations in Africa or America. But the practices of imperialism made the ideological distinctions between Europeans and Africans "come true" to some extent. Only Europeans were permitted to perform the conceptualizing administrative labor that requires the kind of world view the Africanists we examined attribute to Europeans. The conceptualization and administration of complex labor activities was indeed *transferred* from Africa to the imperial nations. Thus the African vs. the European world views are simultaneously ideological constructs of the imperialists, and also "true" reflections of the dichotomized social experience that imperialism went on to create.

Similarly, the emergence of masculine domination among our distant ancestors can be understood as the transfer of the conceptualization and control of women's sexuality, reproduction, and production labor to men—a process intensified and systematized in new ways during the last three centuries in the West. Here, too, the attribution of different natures and world views to women and men presumably occurs originally as an ideological construct by the dominators but subsequently "becomes true" as the control of women's labor is shifted from women to men.

But peoples engaged in struggles against imperialism and masculine dominance are conceptualizing their own labor and experience counter to their rulers' conceptions. It is precisely the disappearance of other-conceptualized labor and experience that permits the emergence of Africanism and feminism. And this disappearance has economic, political, and social origins that lie outside Africanism and feminism. As we have noted, the revolution in birth control, the norm of drawing women into wage labor—and the consequent double-day of work—are conditions that permit women to conceptualize their own labor and experience in new ways. Similarly, the demands of the "internal logic" of capitalism—more consumers, differently skilled labor, and legitimations of both by local, state, and international economic, political, and educational policies—are among the conditions that permit Africans to conceptualize their own labor and experience in new ways. The political dynamics that created "Africans" and "women" in the first place are disappearing, as are the "Africans" and "women" defined originally by the appropriation of the conceptualization of their activity and experience. Those still caught in the economic, political, and intellectual confines of the "feminine" and the "African" are not the movers and shakers of these movements for emancipation. Those who participate in Africanist and feminist political struggles have far more ambiguous race and gender options, respectively, than the Africans and women whose emancipation they would advance. At least among women, it is precisely those whose economic and political options remain only sex-specific, only "traditional," who are most resistant— and for good concrete reasons—to the feminist political agenda.[35]

Thus we should expect differences in cognitive styles and world views from peoples engaged in different kinds of social activities. And we should expect similarities from peoples engaged in similar kinds of social activities. As noted earlier, the kind of account I am suggesting here finds precedents in tendencies within the sociology of knowledge. Examinations of social structure show good reasons why adversarial modes of reasoning are prevalent in one culture and not in another; why instrumental calculation infuses one culture's content and style of thought but not another's. Why is it that the free will vs. determinism dispute does not surface in ancient Greek philosophy but is so central in European thought from the seventeenth century on? Why is it that we can hear nothing about individual rights in ancient Greek thought? Something happened to European bourgeois men's life expectations during the fifteenth to seventeenth centuries to make a focus on individuals and their rights, the effect of the "value-neutral" or impersonal "laws" to which they discovered their bodies were subject, and the power of their wills all crucial problematics for them if they were to understand themselves and the new world they found themselves in.

Was there anything in European women's social experience of the period to lead them to focus on such issues? (Probably yes and no, to read the disputes in history freely.) What about women in traditional nuclear families in the West today? Why should they be expected to hold a world view organized around distinctions between forces outside their control and those within their control, or on problems of adjudicating between the rights of autonomous individuals? What about the social experience of the peoples in the cultures Europe has colonized? Would there be reason for slaves to find interesting the free will vs. determinism dispute, or issues of individual rights? Not much, I am suggesting. For reasons originating in an analysis of social relations, we should expect white, bourgeois European men to have cognitive styles and a world view different from the cognitive styles and world

[35]For my discussion of this issue, see Harding (1983a).

views of those whose daily activities permit the direction of social life by those men.

Developmental Processes.

In the form in which they have been elaborated, the developmental explanations for gender-differing world views favored by the feminist object-relations theorists are thrown into doubt by the overlap of the gendered with the racial dichotomies. Similar processes of producing gender in individuals cross-culturally do not appear powerful enough to produce distinctively masculine and feminine world views cross-culturally—at least not the world views generalized from modern Western gender differences.

Nevertheless, it is possible that object-relations theory can be historicized in illuminating ways. One hint about how to do so is provided by Isaac Balbus. He argues that if we take the intensity of the infant's initial identity with its caretaker (mother) as one cultural variable, and the severity of the infant's separation from that caretaker as another cultural variable, object-relations theory can account for why differing forms of the state arise when they do, and also for culturally differing attitudes toward nature. He points out that some thoroughly misogynous cultures are loath to dominate other cultural groups and/or nature, while some less misogynous cultures regularly engage in the political domination of other groups and the exploitation of natural resources. Balbus is not concerned with issues of racism in his study, and he only begins to explore the anthropological and historical evidence that reveals cultural variations in the intensity of infant identity with the caretaker and the subsequent severity of separation.[36]

Obviously, a great deal of theoretical and empirical work would have to be done to make this intriguing hypothesis capable of explaining how Western men's infantile experience leads to one set of ontologies, ethics, and modes of knowledge-seeking, while the infantile experience of the rest of us tends to produce a different set. However, the core of the "self" we keep for life does appear to be influenced by our prerational experiences as infants—by the opportunities that child-rearing patterns offer us to identify with paternal authority, both as a reaction to and a refuge from initial maternal authority. Thus it

[36]Balbus (1982, esp. ch. 9).

would be foolish to overlook the contributions that a theory of infantile enculturation might make to the "unified field theory" we need.

BACK TO POSTMODERNISM

The logic of the standpoint epistemologies depends on the under-standing that the "master's position" in any set of dominating social relations tends to produce distorted visions of the real regularities and underlying causal tendencies in social relations—including human in-teractions with nature. The feminist standpoint epistemologies argue that because men are in the master's position vis-à-vis women, women's social experience—conceptualized through the lenses of feminist the-ory—can provide the grounds for a less distorted understanding of the world around us.

Euro-American men have had disproportionate responsibility for both racial and gender subjugations and, consequently, doubly dis-torting social experience. Western women have certainly not been innocent of participation in racial subjugations. Women of African descent have had no hand in either. If it is the experience of subjugation that provides the grounding for the most desirable inquiries and knowl-edges, then should not the experience of women of African descent—more generally, women who have suffered from racism—provide the grounding for both African and feminist scientific and epistemological projects, not to mention ethics and politics? Both feminist and Afri-canist political writings often recognize the double oppression of women of color (it may even be triple or quadruple in the presence also of class domination, homophobia, and so on). If the activity of men of African descent and Western women is invisible, more immersed than that of Western men in the concrete and the sensuous, more "me-diating," more unifying of the mental and manual and emotional parts of the self, more estranged from ruling-class conceptual schemes, then surely the activity of women of African descent is even more deeply characterizable in these ways. Does not the internal logic of the stand-point epistemologies demand that the social experiences of women of color provide the starting point for "truer paths" toward belief and social relations undistorted by race and gender loyalties? The uni-versalizing tendencies in each of the successor science and epistemology projects prevent their adherents from drawing these conclusions to which the logic of their own arguments should lead them.

But before we leap to this reconstruction of the grounds of the feminist standpoint epistemologies, let us examine one interesting analysis of the disappearance of the cast of characters required for modernism's epistemological and political dramas—dramas for which the standpoint theorists perhaps only write new dialogue.

Donna Haraway points out that three boundaries fundamental to both the liberal and Marxist versions of humanism have broken down in contemporary social experience. First, "the boundary between human and animal is thoroughly breached. The last beachheads of uniqueness have been polluted if not turned into amusement parks—language, tool use, social behavior, mental events; nothing really convincingly settles the separation of human and animal." Second, the distinction between organism (animal or human) and machine is increasingly difficult to maintain, as contemporary machines "have made thoroughly ambiguous the difference between natural and artificial, mind and body, self-developing and externally-designed, and many other distinctions that used to apply to organisms and machines. Our machines are disturbingly lively, and we ourselves frighteningly inert." Finally, a subset of the second boundary failure is the increasing imprecision of the distinction between physical and nonphysical.[37]

But once the humanist fiction of "man" can no longer be unproblematically naturalized as essentially distinct from animals and machines, or as composed of identifiable components of the physical and the nonphysical—whether these be matter and mind, body and soul, the neurophysical and the social, the endocrinological and the cultural—then the naturalization of its corollary, "woman," is similarly problematized. There is no "*woman*" to whose social experience the feminist empiricist and standpoint justificatory strategies can appeal; there are, instead, *women*: chicanas and latinas, black and white, the "offshore" women in the electronics factories in Korea and those in the Caribbean sex industry. In the concept of "women of color," Haraway sees an identity and a perspective on the world forged out of an "oppositional consciousness" and a politics of solidarity, rather than—as in so much of U.S. feminist theory or the humanist discourses it revises—a naturalized and essentialized identity with a politics of unity.

Furthermore, she finds obstacles to an adequate politics and epistemology for our times in Marxism, in the object-relations theories, and in the radical feminist woman-as-victim-of-masculine-sexuality upon

[37]Haraway (1985, 68–70).

192

which Western feminists lean.[38] All three of those analyses depend on assumptions of the desirability of a return to an original unity of self—a return possible only if we can reunite the fragmented selves created, respectively, by capitalism's alienations, by infantile gendering processes, and (I guess) by the ancient kinship structures Gayle Rubin has named "compulsory heterosexuality."[39] Look at the explanatory benefits that have emerged, Haraway continues, from embracing our "fractured identities" as, say, a Black-feminist, a socialist-feminist, a lesbian-feminist, and so forth. Why not seek a political and epistemological solidarity in our oppositions to the fiction of the naturalized, essentialized, uniquely "human" and to the distortions, perversions, exploitations, and subjugations perpetrated on behalf of this fiction? Why not explore the new possibilities opened up by recognition of the permanent partiality of the feminist point of view?

Haraway's argument would lead to an epistemology that justifies knowledge claims only insofar as they arise from enthusiastic violation of the founding taboos of Western humanism. From this perspective, if there can be "a" feminist standpoint, it can only be whatever emerges from the political struggles of "oppositional consciousnesses"—oppositional precisely to the longing for "one true story" that has been the psychic motor for Western science. Once the Archimedean, transhistorical agent of knowledge is deconstructed into constantly shifting, wavering, recombining, historical groups, then a world that can be understood and navigated with the assistance of Archimedes' map of perfect perspective also disappears. As Flax put the issue, "Perhaps 'reality' can have 'a' structure only from the falsely universalizing perspective of the master. That is, only to the extent that one person or group can dominate the whole, can 'reality' appear to be governed by one set of rules or be constituted by one privileged set of social relationships."[40]

For this feminist postmodernist epistemology, we must begin from diametrically opposite assumptions from those routinely invoked to justify modern science's legitimacy. The greatest resource for would-be "knowers" is our nonessential, nonnaturalizable, fragmented identities and the refusal of the delusion of a return to an "original unity."

---

[38] See MacKinnon (1982) for a paradigm example of the image of woman defined as a victim of masculine sexuality.

[39] Rubin (1975). Radical feminist theory has many strengths, but explanations of the origins of masculine dominance are not among them.

[40] Flax (1986). However, Flax's and Haraway's postmodernisms greatly contrast.

193

But if knowers have come apart, the world has come together. Contrary to the assumption of "a" world out there composed of essential dichotomies, which it is science's job to reconnect through explanation, there are as many interrelated and smoothly connected realities as there are kinds of oppositional consciousness. By giving up the goal of telling "one true story," we embrace instead the permanent partiality of feminist inquiry.[41]

While Haraway develops her account explicitly in opposition to the feminist standpoint strategy, I think it usefully incorporates two key elements of that strategy. First, both depend upon the creation of oppositional consciousnesses, though Haraway's conception of "who's got one" is more concerned with the intersection of race and gender domination than are the standpoint epistemologies. Second, and in contrast to much of mainstream postmodernism, feminist postmodernism, like standpoint approaches, is intensely political.[42] It is here, too, that it reveals the incoherences of much of its "parental" discourse. To refer only to the nonfeminist postmodernist I have found most illuminating, how can we have the "conversation of Mankind"[43] when those who conduct the heretofore politically powerful conversations have such limited tastes, indeed poor tastes, in conversational partners?

In my view, Haraway's analysis is weakened by its still excessive containment within Marxist epistemological assumptions. This can be seen in her not so hidden assumptions that we can, indeed, tell "one true story" about the political economy; that in principle developmental psychologies can make no contributions to our understandings of the regularities and underlying causal tendencies of historical institutions; that we begin to exist as distinctive social persons only when we get our first paycheck or, if we are women, when we first begin adult forms of trading sexual favors for social benefits.

Nevertheless, I think that feminist postmodernism (including Haraway's contributions) offers rich conceptual tools for exploring more

[41]If this all sounds a bit mystical, I ask the reader to think back on how their contemporaries saw the Copernican and Galilean reconstructions of science.

[42]Rorty (1979) hopes that "we" philosophers will be permitted to continue in the "conversations of mankind" as the hegemony of (modern Western) epistemology-centered philosophy declines. (As Tonto said to the Lone Ranger at a moment of crisis in cowboy-and-Indian social relations, "What do you mean 'we,' white man?") See also Paul Feyerabend, *Against Method* (Boston: Schocken Books, 1978), for another example of apolitical postmodernist philosophy. Haraway argues that the feminist postmodernism she explores eschews moralism and vanguard politics. I am skeptical of this claim, and less reluctant to embrace certain kinds of moralisms.

[43]Rorty (1979).

than just the "history of the dead": of "Man," "his culture," "his knowl-edge," and his naturalized and essentialized "woman"—those concepts that humanism's science played such a clear role in constructing and maintaining in their modern forms. Of course, this creates a powerful internal tension: the standpoint epistemologies appear committed to trying to tell the "one true story" about ourselves and the world around us that the postmodernist epistemologies regard as a dangerous fiction. Can the former be sufficiently disengaged from their modernist ances-tors to permit their justification of merely partial but nevertheless "less false" stories?

The problem with relinquishing the successor science projects is that neither feminist theory nor feminist politics stands in a relationship of reciprocity to patriarchal theories and politics. Nor do the former present themselves as proposing merely the respect for difference be-tween men and women that would appropriately characterize the en-visioned postmodernist discourses. The political power of science and its modernist epistemological strategies cannot be left in the hands of those who currently direct public policy, while we theorists dream of a world different from the one that co-opts the "intelligentsia" into the activity of such "harmless" dreaming. Feminists cannot afford to give up the successor science projects; they are central to transferring the power to change social relations from the "haves" to the "have-nots." What else could serve as the epistemological tools for the struggle to change social relations? After all, it is not in the Pentagon or General Motors that one hears of hopes for postmodernism!

On the other hand, we need glimpses in present social relations and understandings, concretely linked to an envisioned future, of the kinds of consciousness many of us are already in fact coming to have. Post-modern tendencies as they appear in feminism provide the best we can manage now for that vision.[44] Feminism cannot afford to give up that vision either, for it is the desired future to which the successor science projects must be in service. This particular apparent tension in feminist thought is simply one we should learn to live with.

But if we Western beneficiaries of humanism's perversions—for many feminist thinkers are such beneficiaries—would join in theorizing a permanently partial science which is *for*, not just *about*, that majority of the members of our species who have fragmented selves and op-

[44]Conversation with Jane Flax has helped me begin to sort out what feminists should want from the varied and complexly related strains of postmodernism.

195

positional consciousness, we need a more robust politics of solidarity than most of us have embraced. White feminists must actively struggle to eliminate the structural racism from which we benefit. As the standpoint theorists point out, an oppositional consciousness is an achievement that requires not only the "science to see beneath the surface of the social relations in which all are forced to participate" but also "the education which can only grow from struggle to change those relations."[45]

[45]Hartsock (1983b, 285).

8 "THE BIRTH OF MODERN SCIENCE" AS A TEXT: INTERNALIST AND EXTERNALIST STORIES

In this chapter and the next, I want to examine in more detail some issues in the history of science. (This is the place where readers who prefer their dramatic story lines uninterrupted by ghostly appearances of the protagonists' ancestors should skip on to the concluding chapter.) Three kinds of science history are incoherent; failure to recognize these incoherences distorts feminist understandings of science, as well as the self-understandings of the sciences we would transform. This historical excursus is intended to make more plausible the epistemological argument of this book, which continues in Chapter 10.

In *The Structure of Scientific Revolutions*, Thomas Kuhn asked what role the history of science should have in the philosophy of science. After all, he said, what is the point of a philosophy of science that cannot account for the processes through which the great breakthroughs in science were achieved? Kuhn argued that the philosophers' rational reconstructions of the history of science, as well as the historians' anecdotes and chronologies, distort the actual processes through which explanations of nature's regularities and underlying causal tendencies have been achieved:

> History, if viewed as a repository for more than anecdote or chronology, could produce a decisive transformation in the image of science by which we are now possessed. That image has previously been drawn, even by scientists themselves, mainly from the study of finished scientific achievements as these are recorded in the classics and, more recently, in the textbooks from which each new scientific generation learns to practice

its trade. . . . This essay attempts to show that we have been misled by them in fundamental ways.

Kuhn believed that a "historiographic revolution in the study of science" was already underway:

> Historians of science have begun to ask new sorts of questions and to trace different, and often less than cumulative, developmental lines for the sciences. Rather than seeking the permanent contributions of an older science to our present vantage, they attempt to display the historical integrity of that science in its own time. They ask, for example, not about the relation of Galileo's views to those of modern science, but rather about the relationship between his views and those of his group, i.e., his teachers, contemporaries, and immediate successors in the sciences.[1]

Kuhn's study directed our attention to the social processes through which inquiry proceeds in science. Along with Jerome Ravetz's analysis of how the institution of science manages its social problems, it set off a veritable renaissance of sociological, historical, and even anthropological studies of science past and present, and created a fruitful disarray in philosophical thinking about the history and present practices of the sciences. However, with but a few exceptions, gender is no more an analytical tool for the post-Kuhnian thinkers than it was for the more traditional observers of science; the usual array of androcentric gaps and distortions appears in these recent studies, too. Even the exceptional gender-sensitive accounts leave unresolved a number of empirical and conceptual issues.[2]

Before looking more closely at the pre-Kuhnians (in this chapter), and the post-Kuhnians (in the next), let us first pull together Kuhn's central arguments with the strands of our earlier comments on the importance of gender as an analytical tool in understanding the history of the enterprise that is the object of so much feminist criticism.

[1]Kuhn (1970, 1–3).

[2]Different kinds of focuses in this literature can be found in Ravetz (1971); Forman (1971); Sohn-Rethel (1978); Mendelsohn, Weingart, and Whitley (1977); Barnes (1977); Bloor (1977); Latour and Woolgar (1979); Knorr-Cetina (1981); Knorr-Cetina and Mulkay (1983). Gender-sensitive accounts include Merchant (1980); Keller (1984); Traweek (1987). These works provide further references to the vast post-Kuhnian literature. See also Griffin (1978), in which a poet put on the feminist conceptual map some of the fundamental issues about science with which we still struggle. Practicing feminist scientists have raised issues about the historical and sociological relationships between gender and science at least since the early 1970s, as noted in earlier chapters.

"The Birth of Modern Science" as a Text

Kuhn showed that activities once regarded as irrelevant or even detrimental to the growth of scientific knowledge were, on the contrary, an integral part of the processes through which hypotheses are developed and legitimated. Perhaps most shocking to his critics was his claim that the armament of conceptual distinctions thought responsible for the great achievements in the history of science were in fact theorized only after the achievements had already been legitimated. Furthermore, Kuhn showed that these conceptual distinctions and methodological directives could not even in principle account for the historical processes they were intended to explain; that is, historians and philosophers of science had credited with the production of scientific revolutions the cognitive structures and inquiry processes that the revolutions only subsequently brought into existence. Perhaps we should understand Kuhn's distinction between revolutionary and normal scientific activity as locating the kind of real distinction reflected in the origins myths that anthropologists report: the processes that generate a mode of human activity or form of social relations are usually different in kind from the activity or relations they generate.

More recent studies of the history and practice of science have pursued the logic of Kuhn's argument in directions he did not take. They have looked not just at the relationship between a scientist's views and those of "his teachers, contemporaries, and immediate successors in the sciences" in order to understand why particular scientific theories and practices developed in the way they did but also at the relationship between scientists' views and those of their predecessors and contemporaries in the whole culture. And they have examined the larger picture of social practices within which particular scientific practices, cognitive structures, and theories gained wide acceptance. That is, where Kuhn's history still tried to present an image of a scientific community that was in significant respects autonomous, subsequent studies have attempted to show the coherence of science with the intellectual and political projects of the cultures within which science takes its place as just one human activity among many.

Nevertheless, most post-Kuhnian social studies of natural science, like their pre-Kuhnian philosophical and historical ancestors, have systematically avoided examining the relationship between gender and science in either its historical or sociological dimensions. Yet if they were to acknowledge that gender is socially constructed and not merely

199

a natural extension of biology, the thoroughly historical understanding their approach calls for would require such an examination. We are now in a position to understand the reasons why this sexist stance produces distortions of history and sociology of science just as it does of any other kind of social science.

In Chapter 2 I identified a series of dogmas of empiricism that have blocked recognition of the desirability of theorizing a positive relationship between gender and science. These dogmas support the defensive belief that science itself should not be examined in the same ways science proposes to examine everything else in the world around us. The post-Kuhnian studies have moved past these dogmas to provide naturalistic and critical interpretive accounts of the history and practice of science. To the extent that they avoid examining the effects of gender identity and behavior, institutionalized gender arrangements, and gender symbolism on the history and practices of science, however, their explanations and interpretations are both incompletely naturalized and distorted.

An earlier chapter also indicated three ways in which a critical understanding of social theory and science is crucial to grasping the effects of gender on the natural sciences. In the first place, the social sciences have tried to imitate the dispassionate, objective methods supposedly responsible for the growth of knowledge in the natural sciences. It may well be that the distorted images of human nature and activity often presented by even the best of social science inquiry are due not merely to the social sciences' different kind of subject matter (conscious, goal-directed actors and their cultures rather than inanimate matter), more complex variables, and relative youth; these inadequacies may also reflect a fundamental problem with the canons of inquiry for the natural sciences. Even in the natural sciences, there appears to be a big gap between the canons of inquiry and social practices. (This was Kuhn's point.) The new social studies of science have not been critical enough; they fail to challenge the pre-Kuhnian assumptions that inquiry canons and the practices of science have been gender-free.

In the second place, natural science is a social phenomenon, created and developed at particular moments in history in particular cultures. Gender (like class, race, and culture) is a variable not only in beliefs about gender differences but also in the most formal structures of beliefs about the boundaries between nature and culture and about the fundamental constituents of socially constructed realities. Thus the

200

formal structures of natural science belief are unlikely to be immune from this kind of gendering. Shouldn't we look at the stories told about the history and rational reconstruction of science as gendered ones?

In the third place, theories about gender and gendered belief, as well as theories about science and its activities, are *social* theories. We all hold lots of "folk beliefs" about what gender is and what science is, but like our culturally inherited beliefs about anything else, these often bear little relationship to the world around and within us. Again, the post-Kuhnian social studies of science, like their empiricist ancestors, in effect treat gender as a biological given rather than as a social construct. Issues raised in the social sciences about periodization, theories of social change, excessive focus on public and dramatic events, the suspicious fit between the conceptual schemes of masculine inquirers and masculine informants, and others that I have discussed must all be explicitly acknowledged before adequate histories and sociologies of science can emerge.

I have also examined the epistemological shifts required to understand knowledge-seeking as a fully social activity—one that will inevitably reflect the conscious and unconscious social commitments of inquirers. From this perspective, it cannot be either merely accidental or irrelevant that most social studies of science, like their empiricist-guided ancestors, are loath to consider the effects on science of gendered identities and behaviors, institutionalized gender arrangements, and gender symbolism.

In examining the incoherent and androcentric meanings of the standard story of the development of modern science, I have chosen to treat concepts and institutions as personages. This approach has several advantages. The processes of birth, growth, and eventual decline of concepts and institutions are in some ways similar to those of individuals. As adults, we usually tell our life stories in ways that obscure to us and our listeners the exact nature of our early painful and partially preconscious struggles. The insights of Freud and Marx have taught us that the accuracy of our autobiographies is limited by what we select as significant, by what we have inadvertently forgotten, by what is too painful to recall, and by what we cannot know about the forces operating in our natural/social surroundings that shaped our early experiences. It is useful to regard the same as true for concepts and institutions such as those of modern science. Histories and sociologies hat are to be critical biographies of a culture—not just self-congratulatory autobiographies—*should* be a "return of the repressed," to borrow Jane Flax's phrase once again;[3] they should reveal to us the ambivalences and gaps in our conscious cultural memories, and their origins in socially repressed histories. Such a quasi-psychoanalytic framework is not, of course, the only one that can critically illuminate the history of science for us but does have its critical virtues, especially when we are interested in the effects of gender on science.

Furthermore, examining science as a personage should appear less bizarre once one notes Kuhn's argument that scientific theories are birthed through kinds of processes different from those responsible for their later growth and development; that the struggles a theory survives in its infancy leave indelible imprints on the character of the mature theory; and that a theory's defenders rewrite its history in a way that often hides the nature of its early struggles. By extending to modern science this same fruitful kind of analysis, we can begin to grasp how modern science—no less than the feudal and other traditional world views with which it contrasts itself—projects onto nature a desired social order. From the beginnings of science, people have looked to nature for information about morals and politics no less than do peoples living in traditional, kin-structured societies.

THE STORY OF SCIENCE'S ORIGINS

All of us grew up on a well-known story about the birth of modern science: who was responsible for the conception, why the labor necessary to bring forth this babe was so difficult, what its birth has meant to three centuries of European and American history, and why the mature personage this babe has become continues to be deserving of massive support in the face of competing demands for public resources. The story is elaborated in mainstream histories and philosophies of science and in scientists' accounts of their lives and their discoveries. It can be found in outline in the opening sections of standard high school and college science texts, and popular accounts of science always explicitly or implicitly allude to it.

Using excerpts from the writings of two highly respected and widely read contemporary historians of science, Thomas S. Kuhn and I. B. Cohen, let's look at the familiar story. Kuhn's *The Copernican Revolution* "aims to display the significance of the Revolution's plurality" and to

[3]Flax (1983).

treat scientific ideas as part of intellectual history: "scientific concepts are ideas, and as such they are the subject of intellectual history."[4] It is especially interesting to examine the Kuhn of 1957 here, since *The Structure of Scientific Revolutions*, which he published only a few years later, is the single work that has done the most to cast doubt on the adequacy of purely intellectual histories. I. B. Cohen's *The Birth of a New Physics* was part of a science study series offering "to students and to the general public the writing of distinguished authors on the most stirring and fundamental topics of physics, from the smallest known particles to the whole universe."[5]

Serious students may find it objectionable to use the images of science presented in popular works as the target for criticism of the history and philosophy of science. I purposely select texts that these historians wrote for general audiences because in them the significance of the standard account is made explicit in a way that would be unnecessary for readers of scholarly accounts, who presumably already understand that a career in science is not just a job but a "calling"—like philosophy, medicine, or the priesthood. Popular histories, unlike scholarly studies, cannot assume that their audiences have been properly socialized into the intellectual arena within which the scholarly studies are produced, or that their audiences make the same assumptions as do scholars about the moral and political significance of science's activities and science's relationship to society; therefore, they make these assumptions explicit. In other words, the moral messages these studies spell out are taken for granted by traditional philosophies and histories of science. If they were not, these scholarly fields could not be so resistant to feminist critiques.

The story traditionally opens with descriptions of the simple and aesthetically pleasing view of the cosmos invented in antiquity. "For most Greek astronomers and philosophers, from the fourth century on," Kuhn tells us, "the earth was a tiny sphere suspended stationary at the geometric center of a much larger rotating sphere which carried the stars. The sun moved in the vast space between the earth and the sphere of the stars. Outside of the outer sphere there was nothing at all—no space, no matter, nothing" (p. 27). Though based in part on everyday observations all of us have made, this picture is not what the universe is really like, for the "two-sphere universe is a product of the

---

[4]Kuhn (1957, vi–viii). Subsequent page references to this work appear in the text.
[5]Cohen (1960, 7). Subsequent page references to this work appear in the text.

human imagination. It is a conceptual scheme, a theory, deriving from observations but simultaneously transcending them" (p. 36). Yet by the late Middle Ages, this simple view of the universe and variants of it had come to exercise an incredible hold over the imagination of Europeans because the religious, moral, and political values of medieval society had been projected onto the ancient astronomy. The astronomy had become a source of information not only about the physical properties of the universe but also about religious, moral, and political values. Hence rational and civilized persons (such as Dante, for instance) turned to this two-sphere astronomical theory "to discover the kinds and even the numbers of the angelic inhabitants of God's spiritual realm" (p. 114).

How different these people were from us!

> No aspect of medieval thought is more difficult to recapture than the symbolism that mirrored the nature and fate of man, the microcosm, in the structure of the universe, which was the macrocosm. Perhaps we can no longer grasp the full religious significance with which this symbolism clothed the Aristotelian spheres. But we can at least avoid dismissing it as mere metaphor or supposing that it had no active role in the Christian's nor [*sic*] astronomical thought. [p. 113]

The reason we should not take this symbolism lightly is that if a purportedly pure description of the universe also carried religious, moral, and political recommendations, then mighty social obstacles would have to be overcome in order to make socially acceptable the revised physical description that Newtonian physics and the subsequent growth of scientific inquiry would in time provide. "When angels become the motive force of epicycles and deferents, the variety of spiritual creatures in God's legion may increase with the complexity of astronomical theory. Astronomy is no longer quite separate from theology. Moving the earth may necessitate moving God's Throne" (p. 114). The emergence of modern science would require a religious, moral, and political revolution.

To understand how the medieval mind could believe what we can now see to be such a simplistic and scientifically problematic view of the universe, Kuhn continues, we should regard this characteristic of the medieval mind as we do the tendency of children and "primitive cultures" to project onto the natural order their own social relations and projects. Like the medieval world view, "the world view of primitive societies and of children tends to be animistic. That is, children and many primitive peoples do not draw the same hard and fast distinction that we do between organic and inorganic nature, between living and lifeless things. The organic realm has a conceptual priority, and the behavior of clouds, fire, and stones tends to be explained in terms of the internal drives and desires that move men and, presumably, animals" (p. 96).

Who could slay this mighty Goliath of a two-sphere view that was keeping European society from scientific and social progress? Armed with but the frail early understandings of the experimental method, and the daring and courage of the hero who fights only for The Right, the great scientists of the fifteenth through seventeenth centuries emerged like guerrilla warriors from the decay and corruption of late medieval society. Beginning with Copernicus's mathematical hypothesis replacing the earth-centered universe with a sun-centered universe, continuing through the theoretical refinements, and accumulating empirical support from inquiries such as Galileo's (with his newly invented telescope), the scientific revolution culminated two centuries later in Newton's universally holding laws of mechanics. Modern science had provided us with the correct one-world view, and this has been the single most significant force for the emergence of pure science, and hence for social progress.

"After Newton's death in 1727," Kuhn explains, "most scientists and educated laymen conceived the universe to be an infinite neutral space inhabited by an infinite number of corpuscles whose motions were governed by a few passive laws like inertia and by a few active principles like gravity. . . . At last the crumbling Aristotelian universe was replaced by a comprehensive and coherent world-view, and a new chapter in man's developing conception of nature was begun" (p. 260). Since the crumbling Aristotelian universe was a moral and political universe as well as a collection of beliefs about nature, breaking its hold over men's minds promised to release morals and politics as well as physics and astronomy from their medieval confines.

The story stresses that the scientific revolution's new method of inquiry would prevent the projection of political interests and values onto the natural order. Modern science, unlike medieval inquiry, seeks knowledge that is free of moral, political, and social values. Truly scientific justification is concerned to establish claims about the regularities of nature and their underlying causal determinants to which all relevantly situated observers, regardless of their personal social or

205

political commitments, can agree. With Galileo's telescope, anyone could see that the heavens were not constituted as the medieval mind believed. Of course, not everyone *wanted* to use the new method to discover the real regularities of nature. "The continuing opposition to the results of telescopic observation is symptomatic of the deeper-seated and longer-lasting opposition to Copernicanism during the Seventeenth Century. Both derived from the same source, a subconscious reluctance to assent in the destruction of a cosmology that for centuries had been the basis of everyday practical and spiritual life" (p. 226).

However, the progress of science since Newton's day confirms that emphases on operationalizing theoretical concepts and obtaining quantitative measures are successful devices for eliminating social values from scientific inquiry. The physical sciences can and do produce statements of fact, and all these statements together provide a picture of nature that is value-free—or at least increasingly value-free. This pure science is a cooperative, consensual enterprise and as such is the most significant testament to man's creativity. Cohen adds:

> Above all, we may see in Newton's work the degree to which science is a collective and a cumulative activity and we may find in it the measure of the influence of an individual genius on the future of cooperative scientific effort. In Newton's achievement we see how science advances by heroic exercises of the imagination rather than by patient collecting and sorting of myriads of individual facts. Who, after studying Newton's magnificent contribution to thought, could deny that pure science exemplifies the creative accomplishment of the human spirit at its pinnacle? [pp. 189–90]

A final excerpt. Though this scientific world view has greatly increased our knowledge of nature and consequently resulted in immense social progress, we must not let down our guard against the ancient enemies: irrational political and religious beliefs. Here is Rudolf Carnap in 1961, describing some beliefs of the members of the Vienna Circle—the group that produced the influential twentieth-century logical empiricist version of this story.

> I think that nearly all of us shared the following three views as a matter of course which hardly needed any discussion. The first is the view that man has no supernatural protectors or enemies and that therefore whatever can be done to improve life is the task of man himself. Second, we had the conviction that mankind is able to change the conditions of life

in such a way that many of the sufferings of today may be avoided and that the external and internal situation of life for the individual, the community, and finally for humanity will be essentially improved. The third is the view that all deliberate action presupposes knowledge of the world, and that the scientific method is the best method of acquiring knowledge, and that therefore science must be regarded as one of the most valuable instruments for the improvement of life.[6]

PROBLEMS WITH THE STORY

Even in such an abbreviated form, this familiar story of the emergence of modern science provides clues to the complex and contradictory meanings science has for modern cultures. We can follow these clues if we treat the story not as a transparent window into history but as an opaque surface that has its own forms and significances.

The History of Science as a Text.
Thinking about this particular account as a story suggests three kinds of texts with which it shares features. First of all, it originates in nonprofessional history. As specialized scholarly disciplines, both the history and the philosophy of science are relatively young; both emerged only in the late nineteenth and early twentieth centuries. Thus the story itself was formed without benefit of the kind of critical (though still incompletely so) concern for the social causes of social phenomena to which historians today direct their attention. Its basic outlines were developed over several centuries, while modern science itself was forming. Its meanings were already part of western European and American society's conscious intellectual inheritance long before the story began to be individually written, only a century ago, by professional historians and philosophers of science. Hence there are good reasons to regard it much as we regard the *Iliad* and the *Odyssey*, the Book of Genesis, or fourth-grade histories of the American Revolution. These are all official and elaborated versions of much older origins stories whose social meanings had already deeply permeated the self-understandings of the people who constructed and listened to them.

Like all other origins stories, this one contains important fragments of natural and social truth. But as texts, these stories inadvertently reveal as much about those who construct them and enjoy listening to

[6]Rudolf Carnap, "Autobiographical Statement," in P. A. Schilpp, ed., *The Philosophy of Rudolf Carnap* (La Salle, Ill.: Open Court, 1963), p. 83.

them as they do about their explicit subject matter. Under the guise of telling people "where we come from," origins stories tell people "who we are." They tell their listeners what human nature is and what goals to strive for in order to live good lives compatible with the "natural order." Internalist histories of science, such as those told by Kuhn and Cohen, have this character: they claim that the discoveries of modern science reflect the pinnacle of human progress, and that the progress science represents is lodged entirely within scientific method. They tell us "who we are": people who use scientific rationality to achieve progress in social life—including, of course, in inquiry.

Second, and more generally, origins myths such as this one are expressions of "folk thought." As noted in Chapter 7, non-Europeans have pointed out that it is not just *their* culture's beliefs that can profitably be studied as folk thought. For many intellectuals and common folk in the West today, this history of science—as well as the elaboration of its central themes by Hume, Locke, Descartes, Kant, and others—is our folk thought. We are not expected to be any more critical of the ability of science to make justifiable claims about the reality hidden beneath appearances than are African villagers of their inherited views of the world around them. The scientific world view was initially adopted as a result of critical thinking (among other reasons), but that is not why most people hold these beliefs today. Critical thinking is not a characteristic of Western thinking just because it is Western; nor is folk thought uniquely characteristic of non-Western thinking.[7]

Finally, the traditional histories and philosophies of science also bear resemblances to autobiographies, especially to the autobiographies of famous and successful people. In this case we are looking at the autobiography of a famous and successful cognitive program and social institution. Autobiographies are selective reports: they reveal what the authors think is important for us to understand about how they came to be the people they are today. Their faithfulness to history is limited by the authors' perceptions of what is significant about their lives; by the failings of memory; by unwillingness and/or inability to recall and report the compromises made and the prices paid for successes, and the painful and often repressed processes through which the authors became adults; and by the degree of ability to understand how crises and achievements experienced as personal events were at least partially destined, shaped by larger social forces. All these limitations on au-

[7]Wiredu (1979); Horton (1967).

tobiographies are also limitations on the familiar accounts of the birth of science. The accounts by scientists themselves, as well as by philosophers and historians, of the famous institution that has advanced their personal and professional lives are limited by these scholars' perceptions of what is significant about the history of science; by the deficiencies of their resource materials; by unwillingness or inability to acknowledge and account for whatever compromises were made in the process of gaining social recognition and social support for their beliefs and practices; and by the inadequate conceptual schemes of the social sciences more generally, which limit our understanding of the forces and desires that have directed social change.

Internal vs. External Histories of Science.
One puzzling aspect of the story of the emergence of modern science is what it claims about the relationship between ideas and social relations. Indeed, this relationship has been the subject of heated dispute among historians and philosophers of science for several decades.

Why do the historians quoted earlier insist that scientific *ideas* have been the single most powerful cause of social progress during the last few centuries, although they recognize that the public *acceptance* of them, which is necessary if these ideas are to be put into practice and have any effect at all on social life, required vast social changes? Of course, we can all acknowledge that individuals come to their beliefs for all kinds of peculiar reasons—including critical thinking about the inadequacies of prevailing ideas. However, changes underway in late feudal and early modern European social life were primarily responsible for the popular acceptance of science's new ways of conceptualizing nature and inquiry. The scientific ideas in turn made more "natural," and therefore morally attractive, the emerging social changes. The answer to my rhetorical question is not hard to find. A complete account would require looking at the mutually causal relations between ideas and social formations—not just at changed ideas as an independent force in history, or at scientific ideas as the mere effects or epiphenomena of independent changes in social formations. But only now are general theories beginning to emerge about the mutual causality exercised by ideas and social relations.

During this century, historians have taken two competing approaches to explaining the rise of modern science. The *internalist* program analyzed the development of modern science "as a cognitive transformation in the history of the endogenous development of intellectual structures"; the *externalist* program sought "the reasons for this transformation in the technical, economic and cultural conditions of society."

> The point of contention between the two programs is that the internal program not only seeks to reconstruct the development of science logically but also to explain it historically. It assumes an independent history of intellectual structures; the development of the forms of knowledge is an independent variable of cultural evolution. The external program, on the other hand, views the social structures and the environment of science not simply as contingent boundary conditions or as a complementary dimension of the development of the logical structures of thought but regards them as constitutive of these.[8]

The account of science we all grew up on is the internalist account. While the externalist account can also be examined as an origins story, as folk thought, and as autobiography within the Marxist discourse, it is the internalist account that has these characteristics within the dominant Enlightenment discourse.

Most historians of science in the early 1960s thought that the legitimacy of intellectual histories was clearly in the ascendant and thus that the debate about the competing programs had come to an end, until the publication of Kuhn's *Structure of Scientific Revolutions* in 1962 reopened the discussion. Since then, a third tendency in social studies of science has attempted to integrate the two programs, though the traditional dispute continues within the two older paradigms.[9] The new syntheses attempt to show how cognitive transformations made specific technical, economic, and cultural changes appear more desirable, and also how historically identifiable social changes led to cognitive changes. The new syntheses have clarified the present practices of science as well, examining the interplay of culture and cognition in contemporary scientific laboratories. We can see the need for synthesis when we look at the internal paradoxes that plague both the earlier approaches to the history of science.

The Internalist's Paradox.

The internalist program assumed that a rational reconstruction of the development of science would simultaneously provide the entire *rele-*

---

[8]Van den Daele (1977, 27).
[9]For examples of continuing discussions within the two older paradigms see, e.g., the journals *Philosophy of Science* and *Telos*.

vant history of science. This is the assumption that motivated the heirs of logical positivism (who usually refer to themselves today as "empiricists") to try to construct ever more nearly perfect logics of justification in the philosophy of science, and still motivates remnants of these attempts among philosophers today. The internalist program tried to produce a rational reconstruction of the development of science for which the history of science would provide no refuting evidence. After all, what point would there be to a logic of justification from which one could draw the conclusion that the historical development of science, and possibly even the reasons why in fact the scientific world view has become dominant, was irrational and *not* logically justified?

However, the development of science is a social phenomenon. When the heirs of logical positivism attempt to prescribe the rules for social inquiry that should guide their own understandings of modern science, internalist assumptions lead them into a peculiar paradox. In the hoary dispute between naturalist vs. intentionalist programs for social science, they explicitly take the naturalist side.[10] Like their intentionalist adversaries, they recognize that human actions are structured not only by the laws governing the behavior of physical matter but also by intentional systems—that is, by culturewide systems of concepts, rules, conventions, and beliefs and by individual systems of perceptions, motives, and goals arrived at within the cultural systems. This fact leads the intentionalist program to insist that this difference in the nature of the subject matter of social inquiry requires deviating from the logics of justification used to explain the causes of natural phenomena. The social inquirer can only interpret for us the significances of regularities for the natives of a particular culture, thereby showing us why actions and institutions that may appear bizarre and irrational to us nevertheless appear rational to the natives. The naturalist position, on the other hand, insists that there is only one explanatory logic for both social and natural phenomena, and that is the logic developed in the natural sciences.

Since development of science is a social phenomenon, how can an

---

[10]Examples of this paradoxical internalist-naturalist position can be found in Ernest Nagel, *The Structure of Science* (New York: Hackett, 1979); and Popper (1959; 1972). Key theorists of "intentionalist" approaches to social inquiry (also called *verstehen*, humanistic, hermeneutical, etc.) are R. G. Collingwood, *The Idea of History* (New York: Oxford University Press, 1956); and Winch (1958). See the discussion of this dispute by Fay and Moon (1977), and Harding (1980).

internalist program in the history of science be defended by those who take a naturalist stance in the philosophy of social science? A *thorough-going* naturalism would have to include in its domain of inquiry the causes for the scientific revolution that lie in the technical, economic, and cultural conditions of the society (including gender conditions). It would have to acknowledge that these "external" causes are not identical with the overt reasons why individual scientists and groups of them found particular hypotheses plausible, let alone with the reasons for our still finding these beliefs plausible. As naturalists are fond of pointing out, in order to explain adequately why people hold the beliefs they do, we need to identify the causes of those beliefs—but these are to be found not in peoples' mental lives but in their environments. Therefore, the internal program in the history and philosophy of science should be seen as defending an intentionalist approach to explaining the development of science alone: every other social phenomenon *but* the development of science requires, they argue, a naturalist explanation!

We can characterize the internalist approach, then, as an incomplete naturalism from the perspective of the philosophy of social science dispute. More strongly, as we argued earlier, the internalist approach protects its position through mystical appeal to an origins myth that forbids bringing scientific rationality to bear on the origins of science itself. The development of science alone is to be understood through the stories scientists and a scientific culture tell about themselves. The rise of science is to be understood through interpretations of the natives' understandings.

This self-understanding of science is incoherent. If science is to be the "measure of all things," it is a conceptual impossibility to measure it by itself. For internalists, the only alternative appears to be to measure science and its claims by standards arising in social relations. But different cultures have wildly differing standards for assessing the adequacy of beliefs and practices, and for the vast majority of cultures these standards are not scientific. On what grounds would we claim modern Western societies more progressive than others if we did not appeal to standards of scientific rationality? For internalists, to abandon assessing the adequacy of social relations by their scientific rationality appears to threaten an absolute relativism: "Man is the measure of all things." Thus they think that the successively more effective adoption in social life of standards of scientific rationality must account for what they take to be the social progress of modern societies. We can understand the problem to which the internalist program is a response, but the restriction of the history of science to the history of an independent cognitive program does not make the problem go away.

The Externalist's Paradox.
The external program in the history of science was developed primarily by Marxists. Boris Hessen, Edgar Zilsel, and J. D. Bernal saw scientific progress as a response to shifts in the economic base of society—the forms of economic production and the social relations that govern them.[11] Taken to one extreme, this political economy of science suggests that human consciousness is entirely a product of such environmental influences as the economic, technological, and political relations of a culture, which "condition" ideas. Thus the scientific beliefs of a particular era—indeed, scientific rationality or even logic—could be regarded as nothing more than cultural expressions of a society's other social arrangements.

If this were true, internalists correctly fear that we would have no transhistorical grounds for arguing that the universe does not rest on the back of a huge tortoise; that the earth really does circle the sun; that Newton's laws are closer to the truth about nature's organization than are Aristotelian views. Furthermore, how would one justify in a nonrelativist way the externalist program and claims themselves? If we have no grounds for judging the desirability of beliefs apart from their coherence with cultural arrangements of which we do or do not approve, why should anyone who is not moved by the Marxist vision of social progress find the externalist accounts plausible?

The externalists themselves appeal to two different arguments to justify their successor science projects in the face of this threat of relativism. On the one hand, they agree with the internalists that the cognitive structures of pure science are transcendental and value-neutral; it is only with the incorporation of science into a bourgeois state that they see the history of the claims, purposes, and uses of science becoming distorted by cultural projects. Hence the core of pure science can be extracted from its bourgeois shell and developed into activities and institutions that are transcendental in the sense that they represent the needs and interests of all humans within the classless society of the future, rather than just the particularistic needs of the bourgeoisie within capitalism. This argument shares with internalism the problem

[11]See, e.g., Hessen (1971); Zilsel (1942); Bernal (1939; 1954).

213

that pure science alone of human artifacts is to be understood through the consciousnesses of its creators rather than through the naturalist explanations that this pure science itself recommends.

On the other hand, the cognitive structures of science are justified as historical emergents accompanying the potential progressiveness of the technologies making possible these cognitive structures. This argument follows from the Marxist claim that the emergence of capitalism was initially a progressive organization of social relations. The amassing of capital made possible the development of technologies that reduce the brute labor necessary to produce food, clothing, and shelter and to satisfy other fundamental needs. And the cognitive structures of science reflect the new social relations of these technologies.[12] But bourgeois life represents an incomplete socialization of labor: many hands are responsible for producing the final products, but the ownership and control of the means of production and of the final products anachronistically remain in private hands. This argument permits us to see some of the social causes of the cognitive structures of science but leaves us wondering whether economic rationality and economic progress really are identical with rationality and progress. To what extent does this story also reflect distorting elements of origins myths, of folk thought, and of autobiography—within the Marxist discourse? Are there not also intentionalist elements within these purportedly naturalist accounts?

So the "internal logics" of both the internalist and externalist programs appear flawed. Beliefs, even scientific beliefs and the cognitive core of scientific beliefs, are not uniquely immune from cultural influences. But neither is it only the economic, technological, and political history of class relations in the fifteenth to seventeenth centuries that we need to understand in order to explain the development of the scientific world view. We want to understand how these social arrangements and others not even dreamed of by the external historians shape human consciousness, but we also want to retain the internalist assumption that not all beliefs are equally good—true, rational, desirable—apart from what we think of the societies that produced them.

The traditional story of the development of modern science holds a deep moral grip on the imaginations and self-images of both intellectuals and common folk in our culture, where scientific rationality more

---

[12]Sohn-Rethel (1978).

and more thoroughly permeates our social relations. Historicizing science requires looking both at and beneath the text: we need to examine the relationship between this story and the actual history of science. But doing so is not an easy task, for the two kinds of history of science ready at hand are both flawed.

Internalist histories cannot explain what now appear to be the obvious influences of modern economic, technological, and political developments upon the formation of the concepts and institutions of science, nor do they leave any epistemological space for looking at the effects of changes and continuities in social relations between the genders upon scientific ideas and practices. And the grounds upon which their own program is to be justified remains mysterious, since its premises conflict with their explicit directive to seek causes that lie outside consciousness.

The externalist program is threatened at every point by relativism. Why should changes in economic, technological, and political arrangements make the new ideas reflecting these arrangements better ideas? Why shouldn't we regard the externalist program itself as simply an epiphenomenon of nineteenth- and twentieth-century social relations destined to be replaced as history moves along? And like the internalists, the externalists also leave no ontological or epistemological space for examining the effects of social relations between the genders on ideas and practices. There are good reasons to think that the "progress" brought by the economic, technological, and political aspects of capitalism was regressive not only for the subsequent victims of bourgeois and imperialist social projects but also for women.

The internalist and externalist studies of science, then, are hopelessly flawed. But with respect to the effects on the growth of scientific knowledge of gendered identities and behaviors, institutional gender arrangements, and gender symbolism, are the new social studies of science any better?

9 PROBLEMS WITH POST-KUHNIAN STORIES

In the preceding chapter we examined ways in which the two standard approaches to the history of science are flawed. The third and more recent approach, the post-Kuhnian social studies of science, offer greater opportunities for using gender as an analytical category, though only a few exceptional accounts actually do so.

One particularly helpful study succeeds in identifying the historical moment in which the political and intellectual foundations for the internalist approach to science were self-consciously formulated. With this kind of account, we can locate the key moment of mythologizing in the lived history of science—the moment when the origins myth, folk thought, and autobiography of science consciously began to take shape—and identify more readily the probable gender dimensions in both the formation of modern science and its mythologies.

Armed with this alternative conceptual framework, we can more easily detect the distinctive modern, Western, and androcentric cosmology that the internalist histories project onto nature, and thus demystify the gap between the progressiveness on behalf of the species that science expresses as its goal and the actual sad history of science's regressive consequences for socially dominated races and classes and for women of all races and classes. The animism that Kuhn considered distinctive of "primitive societies and children" turns out to be characteristic of modern science as well. We can also detect internal inconsistencies in the origins myth which have created central problematics for the contemporary philosophy and history of science.

216

This chapter next examines some problems in contemporary thinking about the role of metaphors in scientific theories—crucial in understanding the continuing power of the metaphors of gender politics so visible in early arguments for the social legitimacy of the scientific method and world view—and, finally, draws attention to some startling continuities between the goals and practices of feminist inquiry today and those of science practitioners prior to the seventeenth-century moment of mythologizing.

THE MOMENT OF MYTHOLOGIZING

If we think about the emergence of modern science as a three-stage process, we can see that the major moment of mythologizing occurred at the beginning of stage three. The first stage was the breakdown of the feudal division of labor that made possible the creation of science's method of experimental observation. The second stage, visible in the New Science Movement in seventeenth-century England, was a political self-consciousness that the features of experimental method seemed to exemplify. The third stage required a further reorganization of social labor, one that compromised the political goals of the New Science Movement and produced the conception of purely instrumentalist, value-neutral science that is increasingly under attack today. Only the third-stage cognitive structure is acknowledged by the rational reconstructions of science; however, the rational reconstructers attribute the third cognitive structure to preceding stages, and they ignore the social structure of science at its third stage, which was responsible for the cognitive structure they recommend.[1]

Of course this chapter provides clues for constructing a "revisionist" history of science only if one takes the internalist histories of science as an accurate record of "what really happened." From the perspective of this study, it is the internalist histories which are revisionist in that they repress consciousness of the origins of science, transforming "what really happened" into the mythologized origins story we examined in Chapter 8.

[1] My argument here echoes those in Kuhn (1970). In the standard "rational reconstructions" of science, the structure of "normal science" is attributed to science's revolutionary moments and then recommended as *the* desirable form of scientific inquiry. Kuhn argued that the "normal science" form did not exist—and could not have existed—during its revolutionary moments, and that a quite different process of inquiry was responsible for the "paradigm shifts" that mark the revolutionary periods.

217

Stage One: The Formation of a New Class.

Edgar Zilsel, a European sociologist of science, argued in the 1930s and 1940s that experimental method cannot be developed in societies that practice slavery.[2] Since experimental method requires *both* educated intelligence *and* the willingness to work with one's hands in designing and manipulating observational technologies, and since in slave societies education must be forbidden to manual laborers lest their ability to read and write provide them with the vision and communicative tools to organize and overthrow their masters, then slaves can never become scientific experimenters. Moreover, in such cultures the distaste for manual labor—the activity considered characteristic of slaves and other menials—is so great among slaveholders that they can never become scientific experimenters, either.

European feudalism was not a slaveholding culture, but the division of labor between the intellectuals and landed aristocracy on the one hand and those who worked the land on the other hand was strong enough to preclude the emergence of scientific experimentation. This division of labor was weakened by the appearance of a new kind of social person whose labor required both educated intelligence and the manipulation of instruments and raw materials. Zilsel identifies six groups of these new workers who appeared during the fourteenth century: artisans, shipbuilders, mariners, miners, foundrymen, and carpenters. Although uneducated in the sense that they were illiterate, they "invented the mariner's compass and guns; they constructed paper mills, wire mills, and stamping mills; they created blast furnaces and . . . introduced machines into mining." Zilsel argues that "they were, no doubt, the real pioneers of empirical observation, experimentation, and causal research."[3]

Zilsel's account allows us to see that it was a violation of the feudal division of labor that permitted experimental observation to emerge and to become a method of inquiry. The technique was not invented by Galileo, Bacon, Harvey, Kepler, and Newton; they only used and refined it. Science's new way of seeing the world developed from the perspective of the new kind of social labor of artisans and the inventors of modern technologies. In turn, the new learning produced by ex-

[2]Zilsel (1942). Interestingly, Zilsel was both a socialist and a member of the Vienna Circle that produced the contemporary version of the dominant positivist philosophy of science.

[3]Zilsel (1942).

perimental observation increased the economic and political importance of this kind of activity and social person. Experimental method became first possible and subsequently important because it approached the world as it could be grasped only from the perspective of a violation, a gap, a free space, in the feudal division of labor.

Stage Two: A New Political Self-Consciousness.
By the seventeenth century the characteristics of experimental observation were among the central features of a self-conscious political movement. The New Science Movement in Puritan England during the 1640s and 1650s—the moment preceding the return of the monarchy—had radical social goals. It was not institutionalized; as Van den Daele explains, it did not yet have "a social role which makes the technical and social elements of science behavior binding," and its cognitive and social programs were not separate.[4] Full of self-confidence and enthusiasm, the various circles of scientists in England saw the political impulse of Puritanism and the struggles of the emerging science as a single progressive dynamic in the shaping of postfeudalism. Science's progressiveness was perceived to lie not in method alone but in its mutually supportive relationship to progressive tendencies in the larger society.

Six aspects of the New Science Movement expressed the integration of science with the progressive political impulses of Puritanism. These are interesting in themselves because they indicate a very different conception of science at its birth than is reported by the standard rational reconstructions. But they also enable us to see important continuities between the feminist empiricist and standpoint reconstructions of science and the early modern scientific impulse.

In the first place, a precondition for the emergence of science was an antiauthoritarian attitude. The revival of learning in the late Middle Ages required opposition against the philosophical authority of Aristotle, Ptolemy, Galen, and other ancients. Paracelsian physicians, alchemists, mystical-hermetic thinkers, and mechanical philosophers had very different projects, but they were united by their shared belief in personal experience as the source of knowledge. This seemed a justifiable belief because experimental observation provided a means through which subjective experience could be reproduced and thus

---

[4]Van den Daele (1977, 28). Subsequent page references to this work (and the authors it cites) appear in the text.

made universal. Furthermore, both the Protestant Reformation and Cartesian rationalism favored a modified evaluation of subjectivity. Experimental observation and the resuscitation of faith in the legitimacy of subjectivity created a new confidence in the individual, which was the intellectual basis for resistance to the authority of the ancients. The same antiauthoritarian stance motivated demands for political emancipation (p. 32).

In the second place, the New Science Movement required the radically novel belief that progress is both desirable and possible. The feudal world view saw change in nature and social life as signifying decay and "corruption." The new learning made possible by science supplied the paradigm for expectations of "an open future, the critical scrutiny of the old and the accumulation of the new" (p. 33).

Third, the part of the science movement inspired by Bacon's vision for the advancement of learning was entirely consistent, "with the democratic, participatory impulse of the Puritan era. It places perception of the senses and real things above rhetoric brilliance and speculative wit. It makes the phenomena of everyday life and the products and procedures of craftmanship the objects of scientific investigation. It emphasizes the role of work as the source of cognition and insists on a clear and plain style and intelligible language in the communication of scientific findings." Such a science would not be the sole possession of an aristocracy for, as Bacon says, "The course I propose for the discovery of sciences is such as leaves but little to the acuteness and strength of wits, but places all wits and understandings nearly on a level." Elsewhere Bacon repeats that "my way of discovering sciences goes far to level men's wits, and leaves but little to individual excellence; because it performs everything by surest rules and demonstrations" (p. 34).

Fourth, the science movement was committed to educational reform. "The philosophy of real, concrete things, the emphasis on experience of the senses, and the valuation of manual labor underlying the New Learning demanded radical alternatives to the traditional schools and universities," and "the reformation of natural knowledge through the experimental method" became symbolic of "a purification of all knowledge from prejudice and corruption" (p. 35). This required the replacement of the ornaments of scholastic learning by learning with public service as its goal.

Fifth, the science movement had a humanitarian orientation. It was concerned to further the public good; in the context of Puritanism, this meant improving the lot of the poor. The benefits of the new learning were to be used to improve nutrition, to create jobs in the towns, to improve health care. Scientific knowledge was to be "for the people" ("science for the people" is Galileo's phrase); it was to be used to redistribute both wealth and knowledge.

Sixth, the new science movement was committed to the unity of theological and philosophical truth. "Although in principle they dissociated religious insight from scientific explanation, the Puritan Baconians always spoke of the 'advancement of piety and learning' in one breath. For them, the progress of science coincided with the truth of the Christian faith and without this was neither true, nor legitimate, nor useful." For example, "for Hermetic chemistry, experience and experiment comprise not only the practical manipulation of objects but they presuppose the intervention of divine illumination without which the secrets of nature cannot be uncovered" (p. 38).

Van den Daele sums up the radical goals of this movement in the following way

> In the scientific movement of Puritan England, we find the idea of experimental natural knowledge embedded in schemes whose claims and norms range far beyond what, to our mind, the concept of positive natural science delimits. Chemical philosophy developed the vision of a mystical and religious knowledge of nature as a Christian alternative to the idle speculation of scholastic philosophy. The Baconian reform movement linked and identified the New Learning with moral, educational, and social aspects. In all the social utopias of the period, the learned societies of the new philosophy . . . are regarded as the basis for a reconstruction of social life. The advancement of science achieved through the co-operation of the philosophers is the means of universal progress, the scientific method is the paradigm of unity through truth. . . . Reflection upon the effects of science is part of, or a condition for, science itself. [p. 38]

Science's new cognitive structures gained support because they were coherent with, one with, the struggle to overthrow the political and intellectual authoritarianism of feudalism. This struggle, in turn, found a powerful justification and instrument for its programs in the new learning that science produced and in science's cognitive structures. The belief that science is inherently emancipatory, which we saw in the standard story of the birth of modern science, emerges only in the projects and meanings of a *prepositive* science, where experimental observation is not yet separable from the historically specific political goals it seemed to advance.

We must here note that there is no doubt more to the story of the New Science Movement; most likely the learned societies were far more self-interested and less populist than Van den Daele's account reveals. It is improbable indeed that the movement's members were not also trying to advance their own interests as a social group through the advancement of the new science. But at least they saw their own interests as coinciding with an emancipatory restructuring of society that shifted power from the "haves" to the "have-nots"; it provides a marked contrast with the ethos of science today.

Stage Three: A New Division of Labor.
Another reorganization of social labor produced the "positive" conception of science as value-free that we have today. In England, faced with the replacement of Puritan progressiveness by absolutist rule and the consequent sanctions against the social programs of the science movement, science opted for a social role that produced behavior-binding norms for those practicing science. Because such a compromise required the separation of its social from its cognitive programs, science's emancipatory potential was thus reduced to its method. It is only at this moment that the internal program in the history of science can properly begin, for it is only here that the intellectual and social goals of science can be separated in practice or concept. Here we have the moment of mythologizing. It becomes possible to see the history of science as part of a purely intellectual history only after this reorganization of labor. Thus, even though the earlier period is the focus of internalist stories, the internal program has an appropriate domain for its history of science only after the separation of science's cognitive program from its social program.

The end of the Puritan Revolution with the Restoration in 1660 also marked the end of the association between science and social, political, or educational reform, and the end of the integration of scientific with religious knowledge. Charles II rescinded the laws of the Interregnum, revoked Cromwell's legal reforms, revised the social policy, abolished the national Puritan Church, and purged the universities of the adherents of experimental natural philosophy (p. 40). The political and social setting of the new science was fundamentally changed.

The chartering of the Royal Society in London in 1662 and of the Académie des Sciences in Paris in 1666 constituted a "decisive step in the social history of science toward the institutionalization of science." The incorporation of these societies "gave birth to institutions which defined scientific standards and began to exercise social control over the observance of such standards. Science was metropolized and hierarchized. . . . Consequently for the first time there developed an infrastructure ensuring the relative continuity of scientific work" (p. 29). However, the price paid for this continuity, social visibility, prestige, and political protection from institutions with rival claims was the abandonment of the social reform goals which had motivated much of the new science in the first place.

> Science continued to be a system which deviated from the dominant culture. It held fast to the rejection of traditional authority, to esteeming manual labour and experience of the senses rather than scholastic erudition, to the demand that its discussions and findings be made public, to universalistic evaluation and to the freedom of communication and exchange. However, the normative implications of science were more or less reduced to the functional conditions of experimental research. The clash with the conservative culture was limited to natural philosophy itself. The demands of the Baconians of the Puritan scientific movement had been release from the restrictions of the *ancien regime*, liberty of religious association, liberty of the press, free trade, reform of monopolistic professional practices, leading to free and socially reoriented medicine, education and law. The virtuosi of the Royal Society for their part were seeking a niche within society, not the reform of that society. [p. 41]

The process of institutionalizing science can usefully be seen as the creation of a new division of labor. The separation of science's cognitive and political programs, and the restriction of scientists to the former, separate those who can legitimately create social/political values from those who can legitimately create facts. The destiny of Modern Man was bifurcated: scientists as scientists were not to meddle in politics; political, economic, and social administrators were not to shape the cognitive direction of scientific inquiry. Such a bifurcation is in practice impossible to maintain, in part because the political realm has the economic power to determine which scientific projects will be funded. More fundamentally, however, individual scientists and the scientific enterprise itself are social artifacts, and the selection and definition of what needs explaining can never be free of social dimensions. Furthermore, the hopes and fears of scientists and the scientific enterprise are projected onto nature and social life, there to provide information

223

about the grounds of morals and politics for those who are legitimated as policy-makers.

By locating the historical compromise that resulted in this division of labor, we can see the ideological components of a key concept in modernism's science: the commitment to value-neutrality. The claim that science is value-neutral was not arrived at through experimental observation (even though only claims so arrived at are supposed to be regarded as justified); it was instead a statement of intent, designed to ensure the practice of science a niche in society rather than the emancipatory reform of that society.

Can it be that gender relations were as unchanging during the thirteenth to seventeenth centuries as their omission from this account suggests? Were there no gender dimensions to the breakdown of the feudal division of labor, to the life opportunities that became available to men and women during the heyday of the New Science Movement, to the effects of the restoration of the monarchy in England? One would expect in times of radical social change, when both formal and informal modes of social control were being challenged, that social relations between the genders would also change, and that these changes and the threats they posed would be reflected in the emerging notion of social progress.

In the few historical studies of science that do examine gender, we can see the beginnings of a story of the emergence of modern science that reveals its direct links with androcentric desires and projects. Furthermore, new histories of other so-called progressive moments in history consistently reveal, first, that women tend to lose status at the moments traditional history marks as progressive. More strongly, democratic social impulses appear systematically to deteriorate women's social powers and opportunities. Thus periodizations of history from the perspective of men's experience cannot capture the events and processes that mark significant changes in women's lives. Second, reactionary moments in history frequently have as a central focus the restoration of whatever forms of social control over women are familiar, and women's sexuality is usually perceived as threatening to social order unless it is policed by men.[5] Finally, theories of social change that are blind to the effects on the men's world of shifts within the women's world—of the effects of reproduction on production—can

[5]See the feminist historians' critical rethinking of so-called "progressive" eras reviewed in Kelly-Gadol (1976).

224

provide only incomplete and distorted understandings even of the men's world.

Thus we can be both appreciative and critical of the new social studies of science of which the Van den Daele account is an example. On the one hand, they enable us to see the historical creation of the ideology of science that would later become positivism—the reduction of science's social value to its method. On the other hand, the systematic avoidance of issues of gender identity and behavior, institutional gender arrangements, and gender symbolism leads to the suspicion that we still have an incomplete and distorted understanding of the emergence of modern science. When gender-sensitive accounts of other moments in history have been produced, our understandings of "progressive eras" have radically shifted. Analyses such as Carolyn Merchant's (examined earlier) suggest that a similar fate awaits traditional understandings of the emergence of modern science.

A MODERN COSMOLOGY

In the last chapter we saw the standard story of the birth of modern science claim that it is only traditional thought that projects human social destinies onto nature. As I quoted Kuhn, "No aspect of medieval thought is more difficult to recapture than the symbolism that mirrored the nature and fate of man, the microcosm, in the structure of the universe, which was the macrocosm."[6] But now we are in a position to see the historical social destinies that modern science, too, projects onto nature. Let us examine what atomism, value-neutrality, and experimental observation would have symbolized for people during the emergence of that science and what they have come to symbolize for us.

Atomism.

In contrast to the organicist view of medieval European and other traditional thought, science presents nature as fundamentally atomistic.[7] Nature is uniform; its most basic units are inert and passive matter. These are distinct and separate from one another; they have no intrinsic, essential connections; they are related only by external forces.

The claim that nature is uniform mirrored the claim of both emerging

[6]Kuhn (1957, 113).
[7]See the discussion of this shift in Merchant (1980).

225

capitalism and the liberal political theory that all men are equal. The new economy's equality was to be advanced through the attempt to make all labor interchangeable. Aristocrats and peasants inherited the kinds of labor they would perform and, consequently, their places in the social order. But for bourgeois man, labor and social status were not to be inherited; nor, eventually, were individuals to be permitted to own unique skill or knowledge. We saw this tendency to make labor "equal" foreshadowed in Bacon's understanding that scientific method "leaves but little to the acuteness and strength of wits, but places all wits and understandings nearly on a level." Capitalism's development would increasingly extract knowledge and skill from individuals and relocate them in machinery and production processes.[8] Technological rationality provided the form within which labor would be reorganized under capitalism. Liberal political theory's equality was to be achieved by enforcing equal protection of the law: the law was to recognize each person as equal to all others in the protection of his rights and enforcement of his duties. So, too, scientific method recognizes each observer as equal to all others (that is, each observer already legitimated as a scientist), and the institutionalization of science provided a social role that made norms for inquiry behavior-binding.

The claim that nature's fundamental units are distinct and separate with no intrinsic connections to one another mirrored the political assertion that individuals are not inextricably bound to the beliefs and practices of the group into which they are born. Social ties are not given by God and nature but constructed by humans; as such, they are changeable by humans. Thus, the social fabric of feudalism was presented as a cultural artifact, not as part of the natural order. Individuals were not inextricably bound into their feudal duties and obligations—one and all could function as separate individuals within emerging modern social relations.

Individuals are naturally inert and passive, atomism claimed. So whatever "motions" of social life one observes in individuals—their behaviors as well as their purposes—are not intrinsic to their natures but products of the external forces of social life. These external forces can be changed so that humans will "move" in different ways.

With three centuries of hindsight, it is difficult to see in this mechanistic atomism a desirable destiny for the classes emerging from

---

[8]Bacon is cited by Van den Daele (1977, 34). See Braverman (1974), and the discussion of the social structure of contemporary science in Chapter 4.

feudalism. Moreover, with a contemporary understanding of the pervasiveness of androcentrism, it is impossible to imagine that the "progressive" movements of the period could have intended to emancipate women as fully as men from the constraints of feudalism. Even if we assume that these so-called progressives sincerely believed that all men are equal, and that all of men's social ties are social products and therefore changeable, they certainly did not believe that the female man is equal to the male man, that the law should reflect this equality, that women's and men's labor is interchangeable, that women's ties to men and children and family are among the ties that regressively bind. No doubt this was in large part because women's natures and activities were not perceived as fully social, as part of the social fabric of feudalism. It was indeed modern *man* for whom the atomism of physics and astronomy projected a social destiny.

Value-Neutrality.

In organicist views, nature has its own values and interests: it is intrinsically purposeful. But the post-Copernicans said that there were primary and secondary properties of nature. Primary qualities were those that produce identical measurements from different observers. As Galileo put the issue, "Whenever I conceive of any material or corporeal substance, I am necessarily constrained to conceive of that substance as bounded and as possessing this or that shape, as large or small in relationship to some other body, as in this or that place during this or that time, as in motion or at rest, as in contact or not in contact with some other body, as being one, many, or few—and by no stretch of imagination can I conceive of any corporeal body apart from these conditions."[9] Secondary qualities were those impermanent ones that produce different measurements from different observers—tastes, odors, colors, "touch," and so forth—and also, though Galileo does not discuss them, the emotional feelings and values produced in the observer by an object.

In other words, only the primary qualities are real properties of nature; secondary qualities are only the subjective, personal properties of individual observers and thus not truly "real." What is real is what can be captured by a value-neutral language using an impersonal—physicalist and quantative—idiom. And this is so even when the objects

---

[9]Galileo, *The Assayer*, trans. A. Danto, quoted in A. Danto and S. Morgenbesser, eds., *Philosophy of Science* (New York: Meridian Books, 1960), p. 27.

being described are themselves persons: the only "real" aspects of humans are their quantifiable and physical aspects. Moreover, there are no privileged authorities in matters of morals or knowledge, no social locations should be valued more highly than others with respect to their occupants' ability to provide the best accounts of nature's regularities and their underlying determinants. "Anyone can see" the way the world is, Galileo said.

The claim that there are no values inherent in nature mirrors the political belief that the distribution and character of the beliefs, interests, and values people hold is the result of social constructs. The impersonal idiom—that is, one peculiar to none of the socially legitimated persons of feudalism—would be the appropriate means of capturing the reality once hidden by the feudally anthropomorphic view of nature held by the Church and State. Thus while subjectivity was more highly evaluated by the science movement, it was only the abstracted, objective properties of nature to which multiple subjectivities would assent that became real. Sensory impressions became less than real, as did politics, morals, and the entire world that emotion and feeling pick out—domains where there appeared to be no abstract, objective truths to which subjectivities would assent. And the denial of privileged authority was, as we have seen earlier, a rejection of the authority of Church and State to have the final word in matters of knowledge and morals.

But insofar as it was only the perspective of bourgeois man which was thought capable of transcending the particularities of social history, this feature of science's cosmology, too, is animistic. Like atomism, it projects the desires of a particular social group in history onto the universe as the natural order.

Method.

Finally, perhaps the most powerful symbol of the new science was method. As we have seen, experimental observation was initially understood as a way of equalizing observers, of making objective generalizations on the basis of subjective experiences. With the institutionalization of science, method began to be understood as norms of inquiry—rules and procedures policed by juries of peers through which disputes could be settled. Here science mirrored the hopes of liberal bourgeois man for an administrative form of ruling, a rule by procedure to replace personal rule by individuals.

"Rule by method" reflects what science's purportedly transhistorical ego grasps; it is the echo in epistemology of nature's "rule by law." In modernism's science the stand-in for this ahistorical ego is the curious phenomenon of the invisible inquirer. Science supposedly speaks in the voice of no particular social individuals; the inquirer is always to make himself as a distinctive social personage invisible to the audiences for the results of his inquiries—and, in the social sciences, to the objects of his scrutiny (which social scientists curiously refer to as the "subjects" of inquiry). In contemporary economic and political realms, this phenomenon appears as the invisible administrator. There are no individuals we can pick out as clearly responsible for economic and political policies; only procedures, techniques, technologies, methods of economic and political organization appear as the agents of the social order. Scientific method and scientific technologies become smarter as the individuals who use these methods and run the technologies become dumber. Rule by method permits knowledge to be transferred from persons to things—from historical individuals to systems and machines, which are also historical creations.

We can understand how method, rule, and impartial law appeared as emancipatory symbols when used to challenge the personal authority of representatives of the medieval church and state. But our contemporary social theory, influenced by psychoanalytic concerns, also reveals the distinctively (Western) masculine desires that are satisfied by the preoccupation with method, rule, and law-governed behavior and activity.[10] Here, too, modern science projects onto nature distinctively (Western) masculine projects and destinies.

Clearly, modern science no less than its predecessors projected onto nature symbols that mirrored the character and fate of the paradigm of modern man. The traditional history and philosophy of science focus on only one set of the meanings these symbols carry—those emancipatory meanings that originated prior to the institutionalization of science. With one eye on the new social studies of science and another on gender theory, we can see in this cosmology the emergence of modern forms of gender totemism. The progressiveness of science is to be found in those of its features that replicate what is thought of in the West as masculine: social autonomy, transcendence of the socially concrete and particular, and epistemic and moral decision-making on the basis of impartial methods, rules, and laws.

[10]See Gilligan (1982); Balbus (1982); Keller (1984).

INTERNAL INCONSISTENCIES IN THE TEXT

We have looked at the characteristics the standard histories of science share with origins stories, with folk thought, and with autobiographies. We have located the moments of creation of two (contradictory) concepts central to these standard histories: the inherent social progressiveness of scientific method, and a "positive" science sharply separated from political, economic, social, and moral goals. And we have looked at the complex but specifically historical information revealed by the cosmology of modernism's science. Histories that are critical biographies rather than merely self-congratulatory autobiographies of cultures return the socially repressed aspects of our self-images to consciousness.

With these historical features of a purportedly transcendental, ahistorical, scientific world view in mind, we can better understand four internal inconsistencies in the standard history of the birth of modern science that have defined the issues for much of twentieth-century philosophy and history of science.

Epistemological Determinism vs. Social Causation.
First, the story stresses epistemological determinism—a form of idealism: the scientific *conception* of nature and inquiry and the information that science produces have been the prime progressive movers in modern social history. As Carnap says: "Science [as a system of knowledge] must be regarded as one of the most valuable instruments for the improvement of life."[11] But we have seen that only the emergence of a new kind of labor made both scientific method and a progressive social order possible. And we saw Kuhn recognize that when "astronomy is no longer quite separate from theology," one precondition for the development of a new science is a new social order.[12]

Are ideas responsible for science's increasing control of nature? Or does a new social order make these ideas plausible and attractive? These two questions identify the limits within which internalist and externalist historians and philosophers of science have framed their problematics; once we begin to place the social activity of science in a more inclusive historical context, we should no longer be happy with such simplistic questions. The history of gender identities and behaviors,

[11]Rudolf Carnap, "Autobiographical Statement," in P. A. Schilpp, ed., *The Philosophy of Rudolf Carnap* (La Salle, Ill.: Open Court, 1963), p. 83.
[12]Kuhn (1957, 114).

institutionalized gender arrangements, and gender symbolism have also played roles in the history of the emergence of modern science, and we need to look at the complex and two-way causal influences between all of a culture's social forms and the kinds of cognitive structures it favors.

The Role of Imagination.

On the one hand the ancient two-world universe was a "product of the human imagination." But on the other hand modern science, too, "advances by heroic exercises of the imagination."[13] Recognition of this fact has shaped the contemporary problematic of distinguishing science from "pseudoscience," particularly from the "prelogical" world views of "primitive societies and children" and from what appear to the heirs of positivism to be the excessively imaginative assumptions of both Marxism and psychoanalysis. What roles (positive and negative) do gender desires play in scientific imagination?

A New Two-World Universe?

The new one-world view was supposedly "comprehensive and co-herent." But what happened to the "internal drives and desires that move men," including those that moved Galileo and Newton? It's true they have been banished from any significant or valued presence *within* the explicit world view of modern science, even though the desirability of the scientific world view is defended through appeals to the social values that science itself supposedly advances: creative imagination, individual initiative, aggressive evidence-gathering, cooperation among peers, consensual decision-making, the production "for everyone" of the results of science, and so on. These scientific values are supposed to be transhistorical human values, in contrast to the particularistic values that individuals acquire because of their historical locations in socially stratified societies.

Is there then another world, invisible to science and to which science is indifferent or perhaps even hostile, where these particularistic values exist? Why should we even agree that scientific values are not also particularistic? What if internal drives and desires exercise strong influences on the world of science from their scientifically invisible location in another world? How then could the one-world view be comprehensive and coherent? Did modern science replace the old

[13]Cohen (1960, 189).

two-world view with a new one in which historically specific social values, interests, and goals are supposed to be kept rigidly separated from the purportedly transcendental scientific values, interests, and goals but in fact are left to range in culturally legitimated forms within the purportedly depoliticized institution of science?

We can see in science's new two-world view the creation of a problematic that has motivated the social sciences' relentless but unsuccessful attempts to duplicate the ontologies of the natural sciences. Only if the reality of the world of emotions, values, and politics can be successfully denied can social science achieve the status and legitimacy attained by the natural sciences. We need to ask what effect modern divisions of labor between the sexes have on science's denial of the reality of the emotional world assigned primarily to women. (Is it an accident that the novel, with its focus on the world denied reality by science and, some have said, a woman's form of literary expression, emerges only in the modern world purportedly ruled by scientific rationality?)

Value-Neutrality vs. Progressive Social Values.
Finally, if it was the social projects of an envisioned emancipatory social order that produced and supported the more objective scientific world view, then is it value-neutrality with which science should be allied? Or does science progress, rather, when it is allied with the political perspectives of those in a society who have the fewest interests in maintaining socially oppressive understandings of the natural order? The contradiction between what Helen Longino has identified as science's "constitutive" value-ladenness and its "contextual" claims to value-neutrality create the problem of defining the sources of science's objectivity.[14]

These four internal inconsistencies are not accidental; they are necessary if the kind of scientific rationality extolled by natural scientists and science enthusiasts is to continue to be perceived as the legitimate mode of sorting beliefs and organizing social relations. Traditional beliefs about scientific rationality are fundamentally incoherent, and the new social studies of science do not succeed in locating some of the sources of the distortions in these beliefs. Our choice is between the insoluble problematics that these contradictions ensure, and a dif-

[14]Helen Longino, "Evidence and Hypothesis," *Philosophy of Science* 46 (March, 1979).

ferent set that promise greater understanding of the science we have and the knowledge-seeking we could have.

THE PROBLEM OF THE ROLE OF METAPHORS

Some historians of science have brought to our attention the persistent presence of metaphors of gender politics in the formal and informal thinking of scientists from the emergence of modern science through the present day.[15] Nature, experimental method, the culture of science, and the relationship between a scientist and his theory have often been conceptualized and defended through gender metaphors and analogies. To defenders of the dogmas of empiricism, this fact has no relevance to what science "really is," but anthropologists would regard as indefensible this dismissal of the significance of metaphor in explanations.

Let us pursue this issue further. Do metaphors function in science when they are no longer explicitly cited? Few traditional philosophers of science think so; for them, the metaphors of gender politics used at the emergence of modern science to make familiar the strange new conceptions of nature and inquiry do not constitute evidence for the claim that science today projects an androcentric cosmology.

"It is true," such a critic will say, "that metaphors linking science and gender politics were present at the emergence of modern science. But what does that have to do with the science we have today? Astronomy, physics, and chemistry, in particular—our models of mature sciences—express their theories and observations in quantitative terms; there is no possibility of metaphoric expression in these highly formalized sciences whose claims are made entirely in mathematical terms. Metaphors from social life may still sometimes appear in the discovery stage of the growth of scientific knowledge, but they quickly disappear through the empirical tests and theoretical refinements that constitute the context of justification. If such metaphors appear in biology and the social sciences today, that is just one more symptom of the immaturity of these fields of inquiry. And if individual scientists may sometimes gratuitously use sexist metaphors in their popular writings, that fact reveals something about *them*, not about the theories they discuss. In no case do metaphors have a legitimate or useful place in mature sciences today."

[15]Merchant (1980); Keller (1984); Jordanova (1980); Bloch and Bloch (1980).

How strong a case does such a critic have? In Chapter 2 I critically examined the faith that science's quantification and its "method" (whatever that is) protect scientific theories from projecting social images and values onto nature. In this chapter I have suggested that such apparently abstract notions as atomism, value-neutrality, and reliance on method themselves reflect historically specific—and probably androcentric—social images of self, other, and community. Let us take a look at the history of the discussion about the nature and role of metaphors in science to grasp the inadequacies in the empiricist view (the "formalists" in the terminology of this literature).

The dispute between the *interactionists* (the defenders of the positive role of metaphor in science) and the *formalists* originated early in this century. In 1914, French physicist and philosopher Pierre Duhem argued that there were two kinds of scientific mind, corresponding to Continental and English temperaments: "on the one hand, the abstract, logical, systematizing, geometric mind typical of Continental physicists, on the other, the visualizing, imaginative, incoherent mind typical of the English." (English Channel hostilities obviously were vivid in Duhem's mind!) Duhem thought that analogies and models might be psychologically useful in formulating theories in the first place but had no lasting significance in the growth of scientific knowledge. He objected to models and analogies on the grounds that they are "superficial and tend to distract the mind from the search for logical order."[16] Other critics have argued that they are also misleading and too often taken literally as explanations of a phenomenon.

In 1920, the English physicist N. R. Campbell raised two objections against these and similar views. In the first place, he said, what we want of a theory is not mere mathematical intelligibility but a kind of intellectual satisfaction. We want to understand in ordinary language just what the regularities and underlying causal determinants of the phenomenon are. Models and analogies are one way to provide this kind of intellectual satisfaction. (See my criticism in Chapter 2 of the view that the mathematical expression of a theory can constitute an explanation in the absence of any guide to how the formulas are to be interpreted and applied to the world around us.) In the second place, the growth of scientific knowledge requires that theories constantly be extended and revised to account for new phenomena. Without the

---

[16]Pierre Duhem, *The Aim and Structure of Physical Theory* (Princeton, N.J.: Princeton University Press, 1954), pt. 1, ch. 4, quoted in Hesse (1966, 1–3).

model's analogy, scientists would have no guide as to which extensions and revisions will be fruitful; it is the analogy of models that permits theories to make predictions in new domains of phenomena. He concludes: "Analogies are not 'aids' to the establishment of theories; they are an utterly essential part of theories, without which theories would be completely valueless and unworthy of the name. It is often suggested that the analogy leads to the formulation of the theory, but that once the theory is formulated the analogy has served its purpose and may be removed or forgotten. Such a suggestion is absolutely false and perniciously misleading."[17]

While Campbell's objections certainly appear significant, it is Duhem's view that most philosophers hold today. Primarily because there appear to be no intelligible models in quantum physics, they have concluded that all other models and analogies to be found in the history and present practice of science are merely psychologically useful comparisons; they have no lasting significance for the nature of the theories that contain them or the intelligibility of the phenomena explained.

Mary Hesse took up the argument in the 1960s. She claimed that metaphors used in science to reconceptualize a domain of inquiry are not merely heuristic devices providing eventually discardable frameworks through which to observe nature; that they do not merely provide comparisons that can be replaced without remainder by an explicit, literal statement of the similarities between the two systems linked by the metaphor. Instead, Hesse points out, as a new theory becomes more widely accepted, the two systems come to seem more and more alike: "They seem to interact and adapt to one another, even to the point of invalidating their original literal descriptions if these are understood in the new, postmetaphoric sense. Men are seen to be more like wolves after the wolf metaphor is used, and wolves seem to be more human. Nature becomes more like a machine in the mechanical philosophy, and actual, concrete machines themselves are seen as if stripped down to their essential qualities of mass in motion" (p. 163).

Moreover, not any old metaphor will do to reconceptualize a given domain of inquiry. To be empirically useful, it must draw on widely understood social meanings.

The suggestion that *any* scientific model can be imposed a priori on *any* explanandum and function fruitfully in its explanation must be resisted.

[17]N. R. Campbell, *Physics, the Elements* (Cambridge, 1920), ch. 6, quoted in Hesse (1966, 4–5). Subsequent page references to the Hesse work appear in the text.

235

Such a view would imply that theoretical models are irrefutable. That this is not the case is sufficiently illustrated by the history of the concept of a heat fluid or the classical wave theory of light. Such examples also indicate that no model even gets off the ground unless some antecedent similarity or analogy is discerned between it and the explanandum. [pp. 161–62]

Hesse argues that these considerations imply

rejection of all views that make metaphor a wholly noncognitive, subjective, emotive, or stylistic use of language. . . . Models, like metaphors, are intended to communicate. If some theorist develops a theory in terms of a model, he does not regard it as a private language but presents it as an ingredient of his theory. Neither can he, nor need he, make literally explicit all the associations of the model he is exploiting; other workers in the field "catch on" to its intended implications—indeed, they sometimes find the theory unsatisfactory just because some implications the model's originator did not investigate, or even think of, turn out to be empirically false. None of this would be possible unless use of the model were intersubjective, part of the commonly understood theoretical language of science, not a private language of the individual theorist. [pp. 164–65]

The role of metaphors in scientific theorizing is rational, even if it violates the rational reconstruction of the growth of scientific knowledge through deduction, "because rationality consists just in the continuous adaptation of our language to our continually expanding world, and metaphor is one of the chief means by which this is accomplished" (pp. 176–77).

One would expect Hesse herself to agree with the feminist critics of science who claim that the character of both scientific inquiry and the gender order have been changed through the use of metaphors of nature as womanly and scientific inquiry as an appropriate activity for consolidating and maintaining masculine gender identity. In a later paper, however, Hesse appears to lose the logic of her own argument.[18] There she asserts that a theory's increasing success at prediction and control of the environment filters out the value judgments that were part of its initial formulation. As a result of its pragmatic success—its

[18]Mary Hesse, "Theory and Value in the Social Sciences," in *Action and Interpretation: Studies in the Philosophy of the Social Sciences*, ed. C. Hookway and P. Petit (New York: Cambridge University Press, 1978).

ability to predict and control the environment—those values become unnecessary and undesired as parts of the theory. According to Hesse, the pragmatic criterion is thus the final arbiter between theories; it will replace the moral, social, and political attractions of a theory with value-free reasons for adopting it.

Metaphors are, of course, one way of expressing value judgments. To say "nature is a machine" in an era of increasing appreciation of the benefits machines can bring is to *recommend* that similar benefits can be gained from nature if it is conceptualized and treated as a machine. To say "nature is rapable"—or, in Bacon's words: "For you have but to follow and as it were hound nature in her wanderings, and you will be able when you like to lead and drive her afterward to the same place again. . . . Neither ought a man to make scruple of entering and penetrating into those holes and corners when the inquisition of truth is his whole object"[19]—is to *recommend* that similar benefits can be gained from nature if it is conceptualized and treated like a woman resisting sexual advances.

What is unclear is why Hesse thinks the social meanings of a theory disappear just because it becomes pragmatically successful. The logic of her earlier defense of an interactive understanding of metaphors would lead to a different line of argument: as a theory becomes pragmatically successful, explicit *appeal* to its original metaphors decreases; such appeal is no longer necessary precisely because of the success of the metaphor in shifting the meanings both of the phenomenon requiring explanation and of the theoretical concepts. That is, as a theory becomes pragmatically successful, its theoretical statements directly present the phenomenon as if the metaphor were literally true. Today we do not need to entice people into thinking of nature as a machine—to point out how nature functions as a clock or a system of levers and pulleys—because Newtonian physics formally conceptualizes nature as isolated parts that impinge on each other only as an effect of external forces, and because we do not need to have drawn to our attention the beneficial aspects of machines. (In fact, science is now busy enticing people into a more modern mechanism: nature is a computer, an information system.) We can summarize this process by saying that in theoretical statements, appeals to metaphors retreat from explicit citation to the assumed form of nature, and to the desired relationships with nature that the theory presents.

[19]Cited in Merchant (1980, 168).

It follows that the appeals to gender politics so evident in the writings of the creators of modern science are no longer necessary, since gender politics has become the form of the scientific enterprise's interactions with the world it studies. At the same time, the form of science legitimates gender politics. As the interactionist theory of metaphors explains, models shift the meanings of phenomena in both domains. That is why scientific activity *can* serve as a way of consolidating and maintaining men's gender identities. Science affirms the unique contributions to culture to be made by transhistorical egos that reflect a reality only of abstract entities; by the administrative mode of interacting with nature and other inquirers; by impersonal and universal forms of communication; and by an ethic of elaborating rules for absolute adjudications of competing rights between socially autonomous—that is, value-free—pieces of evidence. These are exactly the social characteristics necessary to become gendered as a man in our society.

We are now in a position to catch hold of one important germ of truth buried in the value-neutrality thesis. Science can not be made value-neutral in the sense of blocking political values and interests from the conceptual schemes and methodologies that direct scientific inquiry; the important role of metaphors alone would deny this possibility. But science *is* value-neutral in the dangerous epistemological and social sense that it is porous, transparent, to the moral and political meanings that structure its conceptual schemes and methodologies. (Does the construction of such a cultural mechanism itself not reflect certain modern, Western, bourgeois, and masculine values?) The moral and political interests these meanings symbolize go in one end of the scientific enterprise as part of its most abstract constitutive elements, and come out at the other end in the nature and structure of the information that science makes available to public policy. Thus the scientific enterprise is *at its best*, as well as at its worst, a kind of Skinnerian cultural black box. There is and must be constant interaction between science's tendency to reflect social life and social life's tendency to reflect science.

Such considerations return us again to the "problem of the problematic." It is in the context of discovery that scientific problems are identified and just what is problematic about them is defined. The social groups that get to define scientific problematics have already won most of the battle to have their distinctive social experience uniquely legitimated by science. Men—and only white bourgeois men—have consistently monopolized the right to define what counts as a scientific problem. The empiricist canons of inquiry insist that this sphere of discovery is outside its province, that the methodological policing of scientific thought is concerned only with contexts of justification—yet consideration after consideration in this book points to the definition of problematics as a chief culprit in creating the racism, classism, and androcentrism of science to which feminists and others object. In problematizing the selection of scientific problems, feminists expose a phenomenon for which no one will admit responsibility.

A final caution: it is important to understand that scientific theories whose conceptual schemes contain oppressive political metaphors can nevertheless extend our understanding of the regularities of nature and their underlying causal tendencies. After all, the pre-Copernican investigations, shaped by feudal political values, produced a great deal of reliable information about the nature and structure of the universe; modern science did not throw away recognition of all of the regularities of nature charted by earlier investigators. This example also shows that conceptual schemes considered "bad science" at one time in history may nevertheless have brought about great leaps in understanding at an earlier time. Mechanistic metaphors certainly were beneficial to the growth of scientific knowledge in the past. But at what social cost?

We can still ask, for whom was the information useful that science produced through knowledge-seeking guided by gender metaphors? Did it contribute—could it have contributed—to progress for women? Have women, as women, benefited from the "penetration" into virtually all aspects of contemporary social life of forms of scientific rationality that serve to consolidate and maintain bourgeois, Western, masculine identity at the expense of women's abilities to direct their own destinies? If not, then why should anyone mark the birth of modern science as a progressive moment in *human* history?

THE NEW SCIENCE RADICALS?

We have seen that an apparent precondition for the creation of science's new experimental method was the breakdown in the feudal division between mental and manual labor. New kinds of social persons—the artisans, shipbuilders, mariners, miners, foundrymen, and carpenters of the fourteenth century—combined intellectual calculation and reasoning with the manipulation of the physical world in their inventions and technological innovations, thus becoming "the real pi-

239

oneers of empirical observation, experimentation, and causal research."[20] Perhaps a precondition for the emergence of the problematics central to feminist inquiry is a breakdown in the division between emotional labor on one hand and the kinds of mental and manual labor associated with men's work on the other. The problematics and perspectives of feminist inquiry, like those of the modern science pioneers, should then be seen as a consequence of a certain kind of shift in the more general relations of social life. This kind of theorizing can deepen our understanding of my earlier discussions of women scientists as a "contradiction in terms," of the scientific value to be found in the alienated and bifurcated consciousnesses of women inquirers, and of the other grounds claimed for a distinctive feminist epistemological standpoint.[21] The new problematics, concepts, theories, methods, purposes, and results of inquiry emerging from feminist research approach the world from the perspective of a violation, a gap, a free space in the gendered division of labor.

The six traits and goals of the New Science Movement identified earlier in this chapter bear an eerie resemblance to those often stated for feminist inquiry. First, feminism's successor science projects challenge authoritarian attitudes and emphasize personal experience as a source of knowledge; feminism supports the self-confidence of the individual member of subjugated groups heretofore not regarded as social individuals; and political emancipation is central to its purposes of inquiry. These features are not unique to feminism, for they can also be found in a number of manifestations of the twentieth-century "crisis of the West." In its anti authoritarian stance, which is part of a larger field of agitation for social change, feminism replicates the New Science Movement.

Second, just as the New Science Movement required the radical belief that progress was both desirable and possible, the feminist successor science projects require the radical belief that it is possible to redefine political and intellectual progress in ways that reveal the social hierarchies of racism, classism, sexism, and culture-centrism to be not natural, not due to biological differences, but socially created and thus changeable.

Third, the feminist successor science projects echo the participatory impulses of the Puritan era that supported the New Science Move-

<hr>

[20]Zilsel (1942).
[21]See Chapters 3, 6, 7.

ment's activities and beliefs. They emphasize the analysis of social relations between the genders in everyday life, and the role of human activity as the source of cognition. Particularly in the health movement but also in other areas of feminist research, the emphasis is on a style that makes the results of feminist research accessible to all women. It is not the "acuteness and strength of wits" but political struggle and feminist education that produce the new understandings.

Fourth, educational reform has always been as central to feminism as it was to the early science radicals. There is emphasis both on reeducating men to a more realistic and less distorted understanding of women's and men's natures and traditional activities, and on providing women with the kind of knowledge they need to throw off their subjugation. The practical and the emotional are valued more highly in this educational program than abstract knowledge and such "ornaments of modernism" as unquestioning acceptance of distorted conceptual schemes and pronouncements of The Greats in particular disciplines. Central to the program is consciousness-raising. Its effects are felt in the rapid appearance of women's studies programs in universities, high schools, law schools, union halls, street academies, YWCAs, and the like; in the new curriculum and faculty development projects aimed at infusing feminist perspectives into the mainstream curricula and the disciplinary canons; in feminist conferences open to all women; in health, counseling, and alternative technology movements; in the establishment of rape crisis and wife-abuse centers; in a plethora of self-help courses and writings whose topics range from "auto mechanics for women" to "how to get your own divorce"; and in such conceptualizations as Smith's "sociology *for* women." All these attest to the centrality of educational reform directed toward making the practical and emotional knowledge that women have, and need, a central part of everyone's education.

Fifth, like the early science radicalism, feminism has a strong humanitarian orientation. The benefits of the new feminist learning are to be used to improve women's health, to provide economic opportunities for women, to improve child care, to improve public policy, to improve the daily social relations in which we all spend most of our waking hours.

Finally, feminism too seeks a unity of knowledge combining moral and political with empirical understanding. And it seeks to unify knowledge of and by the heart with that which is gained by and about the brain and hand. It sees inquiry as comprising not just the me-

241

chanical observation of nature and others but the intervention of political and moral illumination "without which the secrets of nature cannot be uncovered."

We should note one interesting dissimilarity between the early advocates of a new science and the feminist inquirers: identity of the former with the subjugated of the period was far more voluntaristic than feminists' identity with the condition of women. Women feminists remain women no matter what they do; the New Science Movement's intelligentsia could not be serfs, or the urban or rural poor, no matter what they did. The same kind of dissimilarity appears between Marxist and feminist theorists. Marx and Engels were not, after all, "related to the means of production" in the way that defined the proletariat's condition and consciousness—the condition and consciousness for which they spoke. Thus the "problem of intellectuals" and of "vanguard-ism"—which in different forms made their contributions to deradicalizing both the heirs of the New Science Movement and the twentieth-century left—should be less probable within feminism, or at least less intense than in these other scientific movements for social change. Does it make a difference to the path a revolution takes if the social group that articulates revolution and the group that is to make the revolution are the same?

We should be able to learn from history. One message for feminism is about the deradicalization of our goals and projects, the compromises we make. Since feminist projects are incorporated in societies still fundamentally structured by gender orders, racial orders, class orders, cultural orders, feminism clearly must put central emphasis on practical everyday and long-range efforts to eliminate *all* these forms of domination if it would avoid the unhappy fate of the seventeenth-century New Science Movement. Many individuals and groups have a great deal to lose by the advancement of this radical project, and a great deal to gain by transforming the feminist impulse into just one more element in a nonthreatening pluralistic universe of theoretical discourse, where power relationships remain fundamentally unchanged.

10 VALUABLE TENSIONS AND A NEW "UNITY OF SCIENCE"

It is time to return to our main plot. In this summary chapter, I want to restate in somewhat different ways some of the conclusions I have been drawing from my examination of feminist critiques of science and epistemology.

DILEMMAS AND TENSIONS

One project of this study has been to identify and sort out damaging from valuable incoherences, dilemmas, dissonances, tensions in traditional Western thought and in the feminist critiques. We can tentatively summarize the results by saying that it is the tensions we long to repress, to hide, to ignore that are the dangerous ones. They are the ones to which we give the power to capture and enthrall us, to lead us to actions and justificatory strategies for which we can see no reasonable alternatives. The traditional science discourses are full of such damaging tensions. They encourage us to support coercive scientific claims and practices, and claims *about* science, that are historically mystifying and epistemologically and politically regressive. However, I have been arguing for open acknowledgment, even enthusiastic appreciation, of certain tensions that appear in the feminist critiques. I have been suggesting that these reflect valuable alternative social projects which are in opposition to the coerciveness and regressiveness of modern science.

These considerations lead us to the observation that stable and coherent theories are not always the ones to be most highly desired; there are important understandings to be gained in seeking the social origins of instabilities and incoherences in our thoughts and practices—understandings that we cannot arrive at if we repress recognition of instabilities and tensions in our thought.

The causes of the conceptual instabilities in the feminist science and epistemology critiques are to be found partly in an insufficiently critical focus on the mystifications that modernism perpetrates. They are to be found partly in the plurality of sometimes incompatible theoretical categories we bring from the nonfeminist discourses, modernist or not, to our analyses of gender and science. But they are also to be found in the instability of contemporary social life—in the variety of problems on which our own discourses are meditations.

"Something out there" is changing social relations between races, classes, and cultures as well as between genders—probably quite a few "somethings"—at a pace that outstrips our theorizing. Thus the present situation for feminist analysis is not simply the one Kuhn identifies as a preparadigm stage of inquiry. The social relations that are our object of study, which create and re-create us as agents of knowledge and within which our analytical categories are formed and tested, are themselves in exuberant transformation. Reason, will power, "working over the material," even political struggle will not settle them down now in ways over which feminism should rejoice. It would be historically premature and delusionary for feminism to arrive at a "master theory," at a "normal science" paradigm with conceptual and methodological assumptions that we all think we accept. Feminist analytical categories *should* be unstable at this moment in history. We need to learn how to see our goal for the present moment as a kind of illuminating "riffing" between and over the beats of the various patriarchal theories and our own transformations of them, rather than as a revision of the rhythms of any particular one (Marxism, psychoanalysis, empiricism, hermeneutics, postmodernism . . .) to fit what we think at the moment we want to say. The problem is that we do not know and should not know just what we want to say about a number of conceptual choices with which we are presented—except that the choices themselves create no-win dilemmas for our feminisms. More accurately, the problem is that there is no "we" of feminist theorizing—and recognition of that fact can be a great resource for our politics and knowledge-seeking.[1]

[1]Lugones and Spelman (1983); Hooks (1983); Chapter 7. Additional references to defenses of the virtues of fragmented identities appear in Chapter 7.

244

Valuable Tensions and a New "Unity of Science"

With respect to the gender and science issues, this situation makes the present moment an exciting one in which to live and think, but an inappropriate one in which to conceptualize a definitive overview and critique of these issues. I suggest that central arguments those of us with these concerns have had among ourselves are not resolvable in the terms in which we have been pursuing them. I now think we need to see many of our disputes not as naming issues to be resolved but as pointing to opportunities to come up with better problems. The destabilization of thought has often advanced understanding more effectively than restabilization, and the feminist criticisms of science are a particularly fruitful example of an arena in which the categories of Western thought need destabilization. Though these criticisms began by raising what appeared to be politically contentious but theoretically innocuous questions about discrimination against women in the social structure of science, they have quickly escalated to questioning the most fundamental assumptions of modern Western thought. Thus they challenge the categories within which any solutions to these criticisms might be formulated.

The central strains of feminism present it as a totalizing theory—and reasonably so. Because women and social relations between the genders are everywhere, the subject matter is not containable within any single disciplinary framework or any set of disciplines. All versions of the scientific world view take science to be a totalizing theory; it has been assumed that anything and everything worth understanding can be explained or interpreted within the assumptions of modern science. Yet there is another world hidden from the consciousness of science—the world of emotions, feelings, political values; of the individual and collective unconscious; of social and historical particularity explored by novels, drama, poetry, music, and art—within which we all live most of our waking and dreaming hours under constant threat of its increasing infusion by scientific rationality.[2] Part of the project of feminism is to reveal the relationship between these two worlds—how each shapes and forms the other. Thus in examining the feminist criticisms of science, we have had to examine also the worlds of historical particularity and of psychic repressions and fantasies that constantly intrude, only to be insistently denied in the scientific world view.

[2]Milan Kundera asks if it is an accident that the novel and the hegemony of scientific rationality arose simultaneously: "The Novel and Europe," *New York Review of Books*, July 19, 1984.

Equally important strains in feminism, however, insist that it cannot be a totalizing theory. Once "woman" is deconstructed into "women," and "gender" is recognized to have no fixed referents, feminism itself dissolves as a theory that can reflect the voice of a naturalized or essentialized speaker. It does not dissolve as a fundamental part of our political identities, as a motivation for developing political solidarities—how could it in a world where we can now name the plethora of moral outrages designed exactly to contain us, to coerce us, within each of our culturally specific womanly activities? But because of the historical specificity of sexism's structures, this strain of feminist thought encourages us to cherish and defend our "hyphens"—those theoretical expressions of our multiple struggles.

Instead of fidelity to the assumption of patriarchal discourses that coherent theory is not only a desirable end in itself but also the only reliable guide to desirable action, we can take as our standard of adequate theorizing a fidelity to certain parameters of dissonance with and between the assumptions of these discourses. This approach to theorizing captures the feminist emphasis on contextual thinking and decision-making, and on the processes necessary for gaining understanding in a world not of our own making—that is, where we recognize that we cannot order reality into the forms we might desire. We need to be able to cherish certain kinds of intellectual, political, and psychic discomforts, to see as inappropriate and even dangerous certain kinds of clear solutions to the problems we have been posing.

A number of the central instabilities in the feminist science critiques are created by fundamental tensions between our modernist and postmodernist projects. One is the unnecessary choice between criticizing bad science and criticizing science-as-usual. I have argued that both are necessary projects.

Another is the apparent opposition between constructing a successor science and settling for the different but equally ambitious task of deconstructing the assumptions upon which are grounded anything that resembles the science we know. I have argued that there are good reasons to pursue both projects. Each requires the success of the other, for an adequate successor science will have to be grounded on the resources provided by differences in women's social experiences and emancipatory political projects; and an effective deconstruction of our culture's powerful science requires an equally powerful solidarity *against* regressive and mystifying modernist forces.

A third is the tension between a unitary and a fragmented concep-

tualization of the voice of feminism. I argue for the primacy of fragmented identities but only for those healthy ones constructed on a
solid and nondefensive core identity, and only within a unified opposition, a solidarity against the culturally dominant forces for
unitarianism.

Two additional instabilities have arisen that take us back to the
situation of women in science with which I began my review of the
feminist science critiques.

Affirmative Action: Reform or Revolution?

Although the affirmative action challenges are thought by many to be
the least threatening of the feminist criticisms of science, we have seen
that their solution appears to require vast social changes inside and
outside science. Is it worthwhile expending the immense time, effort,
and agony necessary to carry out the affirmative action struggles when
the root of the problem lies outside science in the organization of
society's gender relations and in the uses and meanings of science more
generally? Well, no and yes. No, because these strategies alone cannot
create equity for women within science; after decades of such activism,
the natural sciences remain a male preserve, and the personal and
political price women pay for "making it" there is often very high.
Yes, because such action does bring small advances, change a few
minds, make a little more space for future generations of women, create
political consciousness and solidarity among the women (and men) who
struggle for equity, and reveal the nature of the beast through its forms
of resistance to "reasonable" demands. No, again, because such piece-
meal actions and small changes of consciousness within one social
institution where relatively few women are located do not begin to
address the political issues crucial for women's day-to-day survival in
the rest of social life. Yes, again, because science is *the* model in our
culture of a supermasculine activity (apart from front-line military
duty, of course); thus even small changes can have a relatively large
effect on social relations between the genders more generally.

In short, we should conceptualize affirmative action strategies within
science as both reformist and revolutionary, primarily because desirable directions for radical change emerge only through our attempts
to make what one might have thought were merely reforms, and because the "mere reforms" have nevertheless created resources for those
radical changes. This paradox reveals the inadequacy of the way the
dichotomy has been conceptualized within the parental Marxist dis-

course from which we have borrowed it. There are crucial differences between making merely cosmetic improvements in an institution and radically changing it, but prescriptions for what women should do within science are not easily guided by this conceptual dualism.[3]

The Scientist as Craft Worker: Anachronism or Resource?

We have seen that the traditional philosophy of science assumes an anachronistic image of the inquirer as a socially isolated genius who selects problems to pursue, formulates hypotheses, devises methods to test the hypotheses, gathers observations, and interprets the results of inquiry. Some of the feminist science critiques also assume this anachronistic image.

On the one hand, we have already discussed the reality that most scientific research today is quite different; these craft modes of producing scientific knowledge were replaced by industrialized modes in the nineteenth century for the natural sciences and by the mid-twentieth century for the vast majority of social science research. Thus the rules and norms posited by the philosophy of science for individual knowledge-seekers are irrelevant to the conduct of contemporary science.

On the other hand, again and again we have seen that it is precisely in the areas of inquiry that do remain organized in craft ways that the most interesting feminist research has appeared.

Since the scientific world view that feminism criticizes was constructed to explain the activity, results, and goals of the craft labor that constituted early scientific activity, and since feminist craft inquiry has produced some of the most valuable new conceptualizations, it looks as if we must simultaneously criticize the misleading image of craft inquiry that serves as a resource for the traditional philosophies of science, *and* develop an appropriate understanding of this way of organizing research in order to illuminate feminist practices. Perhaps the scientific enterprise of today is not scientific at all in the original sense of the term; perhaps only the unity of hand, brain and heart possible in craft labor and the estrangement from the dominant culture such craft practices today require permit the kind of critical perspective essential to achieving understanding. Can it be that feminism and similarly estranged inquiries are the true heirs of the creation of Copernicus, Galileo, and Newton? And that this is true even as fem-

---

[3]See my "Feminism: Reform or Revolution?" in *Philosophical Forum* 5 (no. 1–2) (1973–74); reprinted in Gould and Wartofsky (1980).

inism and other movements of the alienated undermine the epistemology that Hume, Locke, Descartes, and Kant developed to justify to their cultures the new kinds of knowledge that modern science produces?

I have been arguing that we cannot resolve these or the dilemmas we mentioned earlier in the terms in which we have been posing them, and that we should regard these instabilities themselves as valuable resources. If we can learn how to use them, we will be the new heirs of Archimedes as we reinterpret his legacy for our age. During the present decline and fall of what we can usefully think of as the Archimedean era, we can see that his great achievement was not his particular theory about how to create a unified perspective but his inventiveness in creating a new kind of theorizing.

A NEW "UNITY OF SCIENCE"?

If sorting instabilities permits a first commandment of science's ideology to bite the dust, our understanding of the relationship between science and values topples a second. Objectivity is not maximized through value-neutrality—at least not in the way the traditional science discourses have construed these concepts. I have argued that it is only coercive values—racism, classism, sexism—that deteriorate objectivity; it is participatory values—antiracism, anticlassism, antisexism—that decrease the distortions and mystifications in our culture's explanations and understandings. One can think of these participatory values as preconditions, constituents, or a reconception of objectivity, as I have now and then suggested in this study. This strategy colonizes the notion of objectivity, leaving only "objectivism" for the "natives'" meanings of the term.

Apart from decisions about which struggles over rhetorical resources to pursue, this new perspective on the issue brings to the surface a related tension in the feminist science critiques. These critiques appear to make an ironic return to the "unity of science" thesis so beloved of the Vienna Circle, the formulaters of the twentieth-century positivist philosophy of science.

For the Vienna Circle, the sciences formed an ontological and methodological continuum, a hierarchically arranged ordering that placed physics at its pinnacle, followed by the other physical sciences, then the more quantitative and "positive" social sciences (economics and behaviorist psychology were their models) leading the "softer" and qualitatively focused ones (anthropology, sociology, history).[4] The feminist criticisms and reconstructive proposals appear also to assert a unity of science but to reverse the order of the continuum. And this thesis is asserted both as a description of what in fact *is* the case in the sciences and as a prescription for how the sciences *should* be ordered. It has been and should be moral and political beliefs that direct the development of both the intellectual and social structures of science. The problematics, concepts, theories, methodologies, interpretations of experiments, and uses have been and should be selected with moral and political goals in mind, not merely cognitive ones.

But where the Vienna Circle proposed a single methodological and ontological continuum on which to rank the adequacy of different scientific inquiries, this tendency in feminism proposes that a continuum of moral, political, and historical self-consciousness is of primary importance in assessing the adequacy of research practices. Whereas physics ranks high on the former, it certainly falls near the bottom on the latter. Whereas the most illuminating historical, anthropological, and sociological studies may fall low on the former, they rank high on the latter. Thus the paradigm models of objective science are those studies explicitly directed by morally and politically emancipatory interests—that is, by interests in eliminating sexist, racist, classist, and culturally coercive understandings of nature and social life. From the perspective of this second unity-of-science continuum, the more abstract arenas of human thought simply occupy the other end of the continuum; morals and politics appear there, as well, though in their most abstract and least explicit forms. Physics and chemistry, mathematics and logic, bear the fingerprints of their distinctive cultural creators no less than do anthropology and history. A maximally objective science, natural or social, will be one that includes a self-conscious and critical examination of the relationship between the social experience of its creators and the kinds of cognitive structures favored in its inquiry.

To repeat the metaphor I borrowed earlier from behaviorism, science functions primarily as a "black box": whatever the moral and political

---

[4]Yet one more irony: Kuhn (1970)—*The Structure of Scientific Revolutions*—which has proved so crucial in stimulating studies that undermine the notions of science central to the Vienna Circle, was originally published as part of the *International Encyclopedia of Unified Science*, vols. 1–2: *Foundations of the Unity of Science*. The lists of editors and members of the organizational and advisory committees for this series, provided on the back of the title page of Kuhn's study, constitute a useful guide to who was who among the logical positivists. History moves in mysterious ways.

values and interests responsible for selecting problems, theories, methods, and interpretations of research, they reappear at the other end of inquiry as the moral and political universe that science projects as natural and thereby helps to legitimate. In this respect, science is no different from the proverbial description of computers: "junk in; junk out." It is within moral and political discourses that we should expect to find paradigms of rational discourse, not in scientific discourses claiming to have disavowed morals and politics.

This assertion of the priority of moral and political over scientific and epistemological theory and activity makes science and epistemology less important, less central, than they are within the Enlightenment world view. Here again, feminism makes its own important contribution to postmodernism—in this case, to our understanding that epistemology-centered philosophy—and, we may add, science-centered rationality—are only a three-century episode in the history of Western thinking.[5]

When we began theorizing our experiences during the second women's movement a mere decade and a half ago, we knew our task would be a difficult though exciting one. But I doubt that in our wildest dreams we ever imagined we would have to reinvent both science and theorizing itself in order to make sense of women's social experience.

[5]Rorty (1979) puts the point this way.

BIBLIOGRAPHY

Aldrich, Michele L. 1978. "Women in Science." *Signs: Journal of Women in Culture and Society* 4(no. 1).

Andersen, Margaret. 1983. *Thinking about Women.* New York: Macmillan.

Ardener, Edwin. 1972. "Belief and the Problem of Women." In *The Interpretation of Ritual*, ed. J. S. LaFontaine. London: Tavistock. Reprinted in *Perceiving Women*, ed. S. Ardener. London: Malaby Press, 1975.

Arditti, Rita. 1980. "Feminism and Science." In *Science and Liberation*, ed. R. Arditti, P. Brennan, and S. Cavrak. Boston: South End Press.

Arditti, Rita, Pat Brennan, and Steve Cavrak, eds. 1980. *Science and Liberation.* Boston: South End Press.

Arditti, Rita, Renate Duelli-Klein, and Shelly Minden, eds. 1984. *Test-Tube Women: What Future for Motherhood?* Boston: Pandora Press.

Balbus, Isaac. 1982. *Marxism and Domination.* Princeton, N.J.: Princeton University Press.

Barnes, Barry. 1977. *Interests and the Growth of Knowledge.* Boston: Routledge & Kegan Paul.

Bernal, J. D. 1939. *The Social Functions of Science.* New York: Macmillan.

———. 1954. *Science in History.* London: C. A. Watts.

Bernard, Jessie. 1981. *The Female World.* New York: Macmillan.

Bernstein, Richard. 1982. *The Restructuring of Social and Political Theory.* Philadelphia: University of Pennsylvania Press.

Bleier, Ruth. 1979. "Social and Political Bias in Science: An Examination of Animal Studies and Their Generalizations to Human Behavior and Evolution." In *Genes and Gender*, vol. 2, ed. E. Tobach and B. Rosoff. New York: Gordian Press.

———. 1984. *Science and Gender: A Critique of Biology and Its Theories on Women.* New York: Pergamon Press.

Bibliography

Bloch, Maurice, and Jean Bloch. 1980. "Women and the Dialectics of Nature in Eighteenth Century French Thought." In *Nature, Culture and Gender*, ed. C. MacCormack and M. Strathern. Cambridge: Cambridge University Press.

Bloor, David. 1977. *Knowledge and Social Imagery*. London: Routledge & Kegan Paul.

Boch, Gisela. 1983. "Racism and Sexism in Nazi Germany: Motherhood, Compulsory Sterilization, and the State." *Signs: Journal of Women in Culture and Society* 8(no. 3).

Braverman, Harry. 1974. *Labor and Monopoly Capital*. New York: Monthly Review Press.

Brighton Women and Science Group. 1980. *Alice through the Microscope*. London: Virago Press.

Caulfield, Mina Davis. 1974. "Imperialism, the Family, and Cultures of Resistance." *Socialist Revolution* 4(no. 2).

———. 1985. "Sexuality in Human Evolution: What Is 'Natural' in Sex?" *Feminist Studies* 11(no. 2).

Chodorow, Nancy. 1978. *The Reproduction of Mothering*. Berkeley: University of California Press.

Cohen, I. B. 1960. *The Birth of a New Physics*. New York: Doubleday.

Cucchiari, Salvatore. 1981. "The Gender Revolution and the Transition from Bisexual Horde to Patrilocal Band: The Origins of Gender Hierarchy." In *Sexual Meanings: The Cultural Construction of Gender and Sexuality*, ed. S. Ortner and H. Whitehead. New York: Cambridge University Press.

Davis, Angela. 1971. "The Black Woman's Role in the Community of Slaves." *Black Scholar* 2.

de Beauvoir, Simone. 1953. *The Second Sex*, trans. H. M. Parshley. New York: Knopf.

De Lauretis, Teresa. 1984. *Alice Doesn't*. Bloomington: Indiana University Press.

Dinnerstein, Dorothy. 1976. *The Mermaid and the Minotaur: Sexual Arrangements and Human Malaise*. New York: Harper & Row.

Dixon, Vernon. 1976. "World Views and Research Methodology." In *African Philosophy: Assumptions and Paradigms for Research on Black Persons*, ed. L. M. King, V. Dixon, W. W. Nobles. Los Angeles: Fanon Center, Charles R. Drew Postgraduate Medical School.

Ehrenreich, Barbara, and Deirdre English. 1979. *For Her Own Good: 150 Years of Experts' Advice to Women*. New York: Doubleday.

Eisenstein, Zillah. 1981. *The Radical Future of Liberal Feminism*. New York: Longman.

Engels, F. 1972. "Socialism: Utopian and Scientific." In *The Marx and Engels Reader*, ed. R. Tucker. New York: Norton.

Faderman, Lillian. 1981. *Surpassing the Love of Men: Romantic Friendship and Love between Women from the Renaissance to the Present*. New York: Morrow.

253

Bibliography

Fay, Brian, and Donald Moon. 1977. "What Would an Adequate Philosophy of Social Science Look Like?" *Philosophy of Social Science* 7.

Fee, Elizabeth. 1980. "Nineteenth Century Craniology: The Study of the Female Skull." *Bulletin of the History of Medicine* 53.

——. 1981. "Women's Nature and Scientific Objectivity." In *Woman's Nature: Rationalizations of Inequality*, ed. M. Lowe and R. Hubbard. New York: Pergamon Press. Originally appeared as "Is Feminism a Threat to Scientific Objectivity?" in *International Journal of Women's Studies* 4(no. 4).

——. 1984. "Whither Feminist Epistemology of Science." Paper presented to "Beyond the Second Sex" Conference, University of Pennsylvania, April 1984.

Ferguson, Ann. 1981. "Patriarchy, Sexual Identity, and the Sexual Revolution." *Signs: Journal of Women in Culture and Society* 7(no. 1).

Flax, Jane. 1978. "The Conflict between Nurturance and Autonomy in Mother-Daughter Relationships and within Feminism." *Feminist Studies* 4(no. 2).

——. 1983. "Political Philosophy and the Patriarchal Unconscious: A Psychoanalytic Perspective on Epistemology and Metaphysics." In *Discovering Reality: Feminist Perspectives on Epistemology, Metaphysics, Methodology and Philosophy of Science*, ed. S. Harding and M. Hintikka. Dordrecht: Reidel.

——. 1986. "Gender as a Social Problem: In and For Feminist Theory." *American Studies/Amerika Studien*, journal of the German Association for American Studies.

Forman, Paul. 1971. "Weimar Culture, Causality, and Quantum Theory, 1918–1927: Adaptation by German Physicists and Mathematicians to a Hostile Intellectual Environment." *Historical Studies in the Physical Sciences* 3.

Foucault, Michel. 1980. *A History of Sexuality*. Vol. 1: *An Introduction*. New York: Random House.

Gilligan, Carol. 1982. *In a Different Voice: Psychological Theory and Women's Development*. Cambridge, Mass.: Harvard University Press.

Gould, Carol, ed. 1983. *Beyond Domination: New Perspectives on Women and Philosophy*. Totowa, N.J.: Littlefield, Adams.

Gould, Carol, and Marx Wartofsky, eds. 1976. *Women and Philosophy: Towards a Philosophy of Liberation*. New York: Putnam.

Griffin, Susan. 1978. *Woman and Nature: The Roaring inside Her*. New York: Harper & Row.

Gross, Michael, and Mary Beth Averill. 1983. "Evolution and Patriarchal Myths of Scarcity and Competition." In *Discovering Reality: Feminist Perspectives on Epistemology, Metaphysics, Methodology and Philosophy of Science*, ed. S. Harding and M. Hintikka. Dordrecht: Reidel.

Haas, Violet, and Carolyn Perucci, eds. 1984. *Women in Scientific and Engineering Professions*. Ann Arbor: University of Michigan Press.

Hall, Diana Long. 1973–74. "Biology, Sex Hormones and Sexism in the 1920's." *Philosophical Forum* 5.

Haraway, Donna. 1978. "Animal Sociology and a Natural Economy of the Body Politic," pts. 1 and 2. *Signs: Journal of Women in Culture and Society* 4(no. 1).

——. 1981. "In the Beginning Was the Word: The Genesis of Biological Theory." *Signs: Journal of Women in Culture and Society* 6(no. 3).

——. 1983a. "The Contest for Primate Nature: Daughters of Man the Hunter in the Field, 1960–1980." In *The Future of American Democracy*, ed. Mark Kann. Philadelphia: Temple University Press.

——. 1983b. "Signs of Dominance: From a Physiology to a Cybernetics of Primate Society." *Studies in History of Biology* 6.

——. 1985. "A Manifesto for Cyborgs: Science, Technology, and Socialist Feminism in the 1980's." *Socialist Review* 80.

Harding, Sandra, ed. 1976. *Can Theories Be Refuted? Essays on the Duhem-Quine Thesis*. Dordrecht: Reidel.

——. 1978. "Knowledge, Technology and Social Relations." *Journal of Medicine and Philosophy* 3(no. 4).

——. 1979. "Is Equality of Opportunity Democratic?" *Philosophical Forum* 10(no. 2–4).

——. 1980. "The Norms of Social Inquiry and Masculine Experience." In *PSA 1980*, vol. 2, ed. P. D. Asquith and R. N. Giere. East Lansing, Mich.: Philosophy of Science Association.

——. 1981. "What Is the Real Material Base of Patriarchy and Capital?" In *Women and Revolution*, ed. Lydia Sargent. Boston: South End Press.

——. 1982. "Is Gender a Variable in Conceptions of Rationality? A Survey of Issues." *Dialectica* 36(no. 2–3). Reprinted in *Beyond Domination: New Perspectives on Women and Philosophy*, ed. C. Gould. Totowa, N.J.: Rowman & Allenheld.

——. 1983a. "Common Causes: Toward a Reflexive Feminist Theory." *Women and Politics* 3(no. 4).

——. 1983b. "Why Has the Sex-Gender System Become Visible Only Now?" In *Discovering Reality: Feminist Perspectives on Epistemology, Metaphysics, Methodology and Philosophy of Science*, ed. S. Harding and M. Hintikka. Dordrecht: Reidel.

——. 1986. "The Curious Coincidence of Feminine and African Moralities: Challenges for Feminist Theory." In *Women and Morality*, ed. E. Kittay and D. Meyers. Totowa, N.J.: Rowman & Allenheld.

Harding, Sandra, and Merrill Hintikka, eds. 1983. *Discovering Reality: Feminist Perspectives on Epistemology, Metaphysics, Methodology and Philosophy of Science*. Dordrecht: Reidel.

Hartmann, Heidi. 1981a. "The Family as the Locus of Gender, Class, and Political Struggle: The Example of Housework." *Signs: Journal of Women in Culture and Society* 6(no. 3).

——. 1981b. "The Unhappy Marriage of Marxism and Feminism." In *Women and Revolution*, ed. Lydia Sargent. Boston: South End Press.

255

Bibliography

Hartsock, Nancy. 1974. "Political Change: Two Perspectives on Power." *Quest: A Feminist Quarterly* 1(no. 1). Reprinted in *Building Feminist Theory: Essays from Quest*, ed. Charlotte Bunch. New York: Longman, 1981.

———. 1983a. "Difference and Domination in the Women's Movement: The Dialectic of Theory and Practice." In *Class, Race and Sex: Exploring Contradictions, Affirming Connections*, ed. A. Swerdlow and Hannah Lehner. Boston: G. K. Hall.

———. 1983b. "The Feminist Standpoint: Developing the Ground for a Specifically Feminist Historical Materialism." In *Discovering Reality: Feminist Perspectives on Epistemology, Metaphysics, Methodology and Philosophy of Science*, ed. S. Harding and M. Hintikka. Dordrecht: Reidel.

———. 1984. *Money, Sex and Power*. Boston: Northeastern University Press.

Hesse, Mary. 1966. *Models and Analogies in Science*. Notre Dame, Ind.: University of Notre Dame Press.

Hessen, Boris. 1971. *The Economic Roots of Newton's Principia*. New York: Howard Fertig.

Hintikka, Jaakko, and Merrill Hintikka. 1983. "How Can Language Be Sexist?" In *Discovering Reality: Feminist Perspectives on Epistemology, Metaphysics, Methodology and Philosophy of Science*, ed. S. Harding and M. Hintikka. Dordrecht: Reidel.

Hodge, John L., Donald K. Struckmann, and Lynn Dorland Trost. 1975. *Cultural Bases of Racism and Group Oppression*. Berkeley, Calif.: Two Riders Press.

Hooks, Bell. 1981. *Ain't I a Woman?* Boston: South End Press.

———. 1983. *Feminist Theory: From Margin to Center*. Boston: South End Press.

Hornig, Lilli S. 1979. *Climbing the Academic Ladder: Doctoral Women Scientists in Academe*. Washington, D.C.: National Academy of Sciences.

Horton, Robin. 1967. "African Traditional Thought and Western Science," pts. 1 and 2. *Africa* 37.

———. 1973. "Levy-Bruhl, Durkheim and the Scientific Revolution." In *Modes of Thought: Essays on Thinking in Western and Non-Western Societies*, ed. R. Horton and R. Finnegan. London: Faber & Faber.

Hountondji, Paulin. 1983. *African Philosophy: Myth and Reality*. Bloomington: Indiana University Press.

Hubbard, Ruth. 1979. "Have Only Men Evolved?" In *Biological Woman: The Convenient Myth*, ed. R. Hubbard, M. Henifin, and B. Fried. Cambridge, Mass.: Schenkman.

Hubbard, Ruth, M. S. Henifin, and Barbara Fried, eds. 1982. *Biological Woman: The Convenient Myth*. Cambridge, Mass.: Schenkman. Earlier version published 1979 under the title *Women Look at Biology Looking at Women*.

Hull, Gloria, Patricia Bell Scott, and Barbara Smith, eds. 1982. *All the Women Are White, All the Men Are Black, but Some of Us Are Brave: Black Women's Studies*. Old Westbury, Conn.: Feminist Press.

256

Huyssen, Andreas. 1984. "Mapping the Post-Modern." *New German Critique* 33.

Jaggar, Alison. 1983. *Feminist Politics and Human Nature.* Totowa, N.J.: Rowman & Allenheld.

Jameson, Fredric. 1984. "Post-Modernism, or the Cultural Logic of Late Capitalism." *New Left Review* 146.

Jordanova, L. J. 1980. "Natural Facts: A Historical Perspective on Science and Sexuality." In *Nature, Culture and Gender*, ed. C. MacCormack and M. Strathern. New York: Cambridge University Press.

Keita, Lancinay. 1977–78. "African Philosophical Systems: A Rational Reconstruction." *Philosophical Forum* 9(no. 2–3).

Keller, Evelyn Fox. 1978. "Gender and Science." *Psychoanalysis and Contemporary Thought* 1(no. 3). Reprinted in *Discovering Reality: Feminist Perspectives on Epistemology, Metaphysics, Methodology and Philosophy of Science*, ed. S. Harding and M. Hintikka. Dordrecht: Reidel, 1983.

———. 1982. "Feminism and Science." *Signs: Journal of Women in Culture and Society* 7(no. 3).

———. 1983. *A Feeling for the Organism.* San Francisco: Freeman.

———. 1984. *Reflections on Gender and Science.* New Haven, Conn.: Yale University Press.

Keller, Evelyn Fox, and C. Grontkowski. 1983. "The Mind's Eye." In *Discovering Reality: Feminist Perspectives on Epistemology, Metaphysics, Methodology and Philosophy of Science*, ed. S. Harding and M. Hintikka. Dordrecht: Reidel.

Kelly-Gadol, Joan. 1976. "The Social Relation of the Sexes: Methodological Implications of Women's History." *Signs: Journal of Women in Culture and Society* 1(no. 4).

Kittay, Eva, and Diana Meyers. 1986. *Women and Morality.* Totowa, N.J.: Rowman & Allenheld.

Kline, Morris. 1980. *Mathematics: The Loss of Certainty.* New York: Oxford.

Knorr-Cetina, Karin. 1981. *The Manufacture of Knowledge.* Oxford: Pergamon.

Knorr-Cetina, Karin, and Michael Mulkay, eds. 1983. *Science Observed: Perspectives on the Social Study of Science.* Beverly Hills, Calif.: Sage.

Kolakowski, Leszek. 1969. *The Alienation of Reason: A History of Positivist Thought*, trans. N. Guterman. Garden City, N.Y.: Anchor Books.

Kuhn, Annette. 1982. *Women's Pictures.* Boston: Routledge & Kegan Paul.

Kuhn, Thomas S. 1957. *The Copernican Revolution.* Cambridge, Mass.: Harvard University Press.

———. 1970. *The Structure of Scientific Revolutions*, 2nd ed. Chicago: University of Chicago Press.

Lakatos, Imre, and Alan Musgrave, eds. 1970. *Criticism and the Growth of Knowledge.* New York: Cambridge University Press.

Latour, Bruno, and Steve Woolgar. 1979. *Laboratory Life: The Social Construction of Scientific Facts.* Beverly Hills, Calif.: Sage.

Bibliography

Leacock, Eleanor B. 1982. *Myths of Male Dominance*. New York: Monthly Review Press.

Leibowitz, Lila. 1978. *Females, Males, Families: A Biosocial Approach*. N. Scituate, Mass.: Duxbury Press.

Leiss, William. 1972. *The Domination of Nature*. Boston: Beacon Press.

Longino, Helen, and Ruth Doell. 1983. "Body, Bias, and Behavior: A Comparative Analysis of Reasoning in Two Areas of Biological Science." *Signs: Journal of Women in Culture and Society* 9(no. 2).

Lowe, Marian, and Ruth Hubbard, eds. 1983. *Woman's Nature: Rationalizations of Inequality*. New York: Pergamon Press.

Lugones, Maria C., and Elizabeth V. Spelman. 1983. "Have We Got a Theory For You! Feminist Theory, Cultural Imperialism and the Demand for the Women's Voice." *Hypatia: A Journal of Feminist Philosophy* (special issue of *Women's Studies International Forum*) 6(no. 6).

Lukacs, G. 1971. *History and Class Consciousness*. Cambridge, Mass.: MIT Press.

Lyotard, Jean-François. 1984. *The Post-Modern Condition*. Minneapolis: University of Minnesota Press.

MacCormack, Carol, and Marilyn Strathern, eds. 1980. *Nature, Culture and Gender*. Cambridge: Cambridge University Press.

McIntosh, Peggy. 1983. "Interactive Phases of Curricular Revision: A Feminist Perspective." Working paper no. 124. Wellesley, Mass.: Wellesley College Center for Research on Women.

MacKinnon, Catherine. 1982. "Feminism, Marxism, Method, and the State: An Agenda for Theory." *Signs: Journal of Women in Culture and Society* 7(no. 3).

Marks, Elaine, and Isabelle de Courtivron, eds. 1980. *New French Feminisms*. Amherst: University of Massachusetts Press.

Marx, Karl. 1964. *Economic and Philosophic Manuscripts of 1844*, ed. Dirk Struik. New York: International Publishers.

———. 1970. *The German Ideology*, ed. C. J. Arthur. New York: International Publishers.

Mendelsohn, Everett, Peter Weingart, and Richard Whitley, eds. 1977. *The Social Production of Scientific Knowledge*. Dordrecht: Reidel.

Merchant, Carolyn. 1980. *The Death of Nature: Women, Ecology and the Scientific Revolution*. New York: Harper & Row.

Millman, Marcia, and Rosabeth Moss Kanter, eds. 1975. *Another Voice: Feminist Perspectives on Social Life and Social Science*. New York: Anchor Books.

Moulton, Janice. 1983. "A Paradigm of Philosophy: The Adversary Method." In *Discovering Reality: Feminist Perspectives on Epistemology, Metaphysics, Methodology and Philosophy of Science*, ed. S. Harding and M. Hintikka. Dordrecht: Reidel.

National Science Foundation. 1982. *Women and Minorities in Science and Engineering*. Washington, D.C.: NSF.

Needham, Joseph. 1976. "History and Human Values: A Chinese Perspective for World Science and Technology." In *Ideology of/in the Natural Sciences*, ed. H. Rose and S. Rose. Cambridge, Mass.: Schenkman.

O'Brien, Mary. 1981. *The Politics of Reproduction*. New York: Routledge & Kegan Paul.

Ortner, Sherry. 1974. "Is Female to Male as Nature Is to Culture?" In *Women, Culture and Society*, ed. M. Z. Rosaldo and L. Lamphere. Stanford, Calif.: Stanford University Press.

Ortner, Sherry, and Harriet Whitehead, eds. 1981. *Sexual Meanings: The Cultural Construction of Gender and Sexuality*. New York: Cambridge University Press.

Owens, Craig. 1983. "The Discourse of Others: Feminists and Post Modernism." In *The Anti-Aesthetic: Essays in Post Modern Culture*, ed. Hal Foster. Port Townsend, Wash.: Bay Press.

Popper, Karl. 1959. *The Logic of Scientific Discovery*. New York: Basic Books.

———. 1972. *Conjectures and Refutations: The Growth of Scientific Knowledge*. 4th ed. rev. London: Routledge & Kegan Paul.

Quine, Willard Van Orman. 1953. "Two Dogmas of Empiricism." In *From a Logical Point of View*. Cambridge, Mass.: Harvard University Press.

Ravetz, Jerome R. 1971. *Scientific Knowledge and Its Social Problems*. New York: Oxford University Press.

Roberts, H., ed. 1981. *Doing Feminist Research*. Boston: Routledge & Kegan Paul.

Rorty, Richard. 1979. *Philosophy and the Mirror of Nature*. Princeton, N.J.: Princeton University Press.

Rose, Hilary. 1983. "Hand, Brain and Heart: A Feminist Epistemology for the Natural Sciences." *Signs: Journal of Women in Culture and Society* 9(no. 1).

———. 1984. "Is a Feminist Science Possible?" Paper presented to MIT Women's Studies Program, April 1984.

Rose, Hilary, and Steven Rose, eds. 1976. *Ideology of/in the Natural Sciences*. Cambridge, Mass.: Schenkman.

Rossiter, Margaret. 1982a. "Fair Enough?" *Isis* 72.

———. 1982b. *Women Scientists in America: Struggles and Strategies to 1940*. Baltimore, Md.: Johns Hopkins University Press.

Rothschild, Joan. 1983. *Machina ex Dea: Feminist Perspectives on Technology*. New York: Pergamon Press.

Rubin, Gayle. 1975. "The Traffic in Women: Notes on the 'Political Economy' of Sex." In *Toward an Anthropology of Women*, ed. Rayna Rapp Reiter. New York: Monthly Review Press.

Ruddick, Sara. 1980. "Maternal Thinking." *Feminist Studies* 6(no. 2).

Sayers, Janet. 1982. *Biological Politics: Feminist and Anti-Feminist Perspectives*. New York: Tavistock Publications.

Bibliography

Signs: Journal of Women in Culture and Society. 1975 et seq. Review essays of feminist research published in every issue.

——. 1978. Special issue on women and science, 4(no. 1).

——. 1981. Special issue on French feminism, 7(no. 1).

Smith, Dorothy. 1974. "Women's Perspective as a Radical Critique of Sociology." *Sociological Inquiry* 44.

——. 1977. "Some Implications of a Sociology for Women." In *Woman in a Man-Made World: A Socioeconomic Handbook*, ed. N. Glazer and H. Waehrer. Chicago: Rand-McNally.

——. 1979. "A Sociology For Women." In *The Prism of Sex: Essays in the Sociology of Knowledge*, ed. J. Sherman and E. T. Beck. Madison: University of Wisconsin Press.

——. 1981. "The Experienced World as Problematic: A Feminist Method." Sorokin Lecture no. 12. Saskatoon: University of Saskatchewan.

Sohn-Rethel, Alfred. 1978. *Intellectual and Manual Labor*. London: Macmillan.

Stacey, Judith, and Barrie Thorne. 1985. "The Missing Feminist Revolution in Sociology." *Social Problems*, 32.

Stehelin, Liliane. 1976. "Sciences, Women and Ideology." In *Ideology of/in the Natural Sciences*, ed. Hilary Rose and Steven Rose. Cambridge, Mass.: Schenkman.

Tanner, Nancy. 1981. *On Becoming Human*. New York: Cambridge University Press.

Tanner, Nancy, and Adrienne Zihlman. 1976. "Women in Evolution," pt. 1: "Innovation and Selection in Human Origins." *Signs: Journal of Women in Culture and Society* 1(no. 3).

Tobach, Ethel, and Betty Rosoff, eds. 1978, 1979, 1981, 1984. *Genes and Gender*, vols. 1–4. New York: Gordian Press.

Traweek, Sharon. 1987. *Particle Physics Culture: Buying Time and Taking Space.* Forthcoming.

Van den Daele, W. 1977. "The Social Construction of Science." In *The Social Production of Scientific Knowledge*, ed. E. Mendelsohn, P. Weingart, R. Whitley. Dordrecht: Reidel.

Walkowitz, Judith. 1983. *Prostitution and Victorian Society: Women, Class and the State*. New York: Cambridge University Press.

Walsh, Mary Roth. 1977. *Doctor Wanted, No Women Need Apply: Sexual Barriers in the Medical Profession, 1835–1975*. New Haven, Conn.: Yale University Press.

Watson, James. 1969. *The Double Helix*. New York: New American Library.

Weeks, Jeffrey. 1981. *Sex, Politics and Society: The Regulation of Sexuality since 1800*. New York: Longman.

Wellman, David. 1977. *Portraits of White Racism*. New York: Cambridge University Press.

Westkott, Marcia. 1979. "Feminist Criticism of the Social Sciences." *Harvard Educational Review* 49.

Winch, Peter. 1958. *The Idea of a Social Science and Its Relation to Philosophy.* New York: Humanities Press.

Wiredu, J. E. 1979. "How Not to Compare African Thought with Western Thought." In *African Philosophy: An Introduction*, 2nd ed., ed. Richard A. Wright. Washington D.C.: University Press of America.

Zihlman, Adrienne. 1978. "Women in Evolution," pt. 2: "Subsistence and Social Organization among Early Hominids." *Signs: Journal of Women in Culture and Society* 4(no. 1).

———. 1981. "Women as Shapers of the Human Adaptation." In *Woman the Gatherer*, ed. Frances Dahlberg. New Haven, Conn.: Yale University Press.

———. 1985. "Gathering Stories for Hunting Human Nature." *Feminist Studies* 11:2.

Zilsel, Edgar. 1942. "The Sociological Roots of Science." *American Journal of Sociology* 47.

Zimmerman, Jan, ed. 1983. *The Technological Woman: Interfacing with Tomorrow.* New York: Praeger.

Zimmerman, Bill, et al. 1980. "Science for the People." In *Science and Liberation*, ed. R. Arditti, P. Brennan, and S. Cavrak. Boston: South End Press.

INDEX

Index

Index

Gender difference: faulty assumptions regarding, 18, 52–53, 89–90, 176; gender complementarity and, 130; social creation of, 17–18, 134, 173, 174–175. *See also* Gender dichotomy (femininity vs. masculinity); Sex difference

Gender identity: and "degendering," 53, 146; effect of, 33, 47, 201; of infants, *see* Children; masculine, 39, 54, 125, 152, 236, 238; masculine, "threatened," *see* Males; relationship of, with gender symbolism and division of labor, *see* Gender symbolism/totemism; sex difference relation to, 17–18, 134. *See also* Sex/gender system

Gender politics, metaphors of. *See* Gender symbolism/totemism

Gender roles: biological view of, 92–93; in the Enlightenment, 117, 118; stereotyped, 60–61, 63, 72, 82, 86, 89, 108, 118

Gender structure. *See* Labor, division of (gender structure)

Gender symbolism/totemism: and androcentric bias, 104, 135; effect of, 33, 39, 201; and metaphors (historical and contemporary) of gender politics, 17, 23–24, 112–126, 145, 176–177, 217, 233–239; in physics, 47; relationship of, with division of labor and individual gender, 52, 53–54, 56, 59, 81, 111; as "threat," 67–68

Gender theorization, 17, 19–24; inadequacy of, 57; obstacles to, 30–36, 37, 51–52, 200; questions raised by, 55–56

Gilligan, Carol, 31n1, 133, 167n5, 171n7, 184n30, 229n10

Gödel, Kurt, 49n14

Gould, Carol, 123n14

Greeks, 49–50, 124, 136, 146, 189, 203

Griffin, Susan, 23n10, 119n6, 123n14, 136n1, 198n2

Gross, Michael, 22n9, 92n13

Guntrip, Harry, 131

Haas, Violet, 63n4

Hall, Diana Long, 119n6

Haraway, Donna, 20n4, 24n12, 28n16, 55n21, 92n13, 96n16, 102n27, 103, 127n20, 136–137, 138, 164n2, 178n18, 192–194

Hartmann, Heidi, 75n21, 75n22, 87

Hartsock, Nancy, 24n12, 26n14, 55n20, 76n23, 143n10, 146, 147–148, 149–151, 156, 158, 196n45

Harvey, William, 218

Hegel, G. W. F., 26, 123, 156, 158

Henifin, M. S., 22n9, 92n13, 125n17, 136n1, 176n16

Hermaphroditism, 127

Hermeneutics (intentionalism), 35, 244; "naturalists" vs., 76–77, 83–84, 211–212

Hesse, Mary, 113n2, 234n16, 235–237

Hessen, Boris, 213

Hiernaux, Jean, 181n22

Hintikka, Jaakko, 51

Hintikka, Merrill, 51, 136n1, 171n7, 178n18

History: African, 173; biased periodization of, 66, 80, 201, 224; images of gender in, 30, 31, 112–119, 173, 174; of science, *see* Science

Hobbes, Thomas, 133, 154

Hochschild, Arlie, 86n3

Hodge, John L., 167n6

Homosexuality, 128–129, 176, 180n21; and homophobia, 16, 21, 22

Hooks, Bell, 166n4, 244n1

Hornig, Lilli S., 63n4, 64n6

Horton, Robin, 174n13, 182n25, 183, 208n7

Hountondji, Paulin, 167n6, 173n9, 174n13, 182n25

Hubbard, Ruth, 22n9, 92n13, 94, 97n20, 101nn24–26, 124–125, 136n1, 176n16

Hudson, L., 121n9

Hull, Gloria, 179n19

Hume, David, 140, 173n9, 208, 249

Hunter-gatherer societies. *See* Labor, division of (gender structure)

"Hyphenization" ("fractured identity"), 26, 28, 163–164, 177, 193, 244n1, 246, 247

Identity. *See* Gender identity; "Hyphenization" ("fractured identity"); Research

Imperialism, 17, 22, 160, 185; colonialism and, 16, 25, 61, 123, 167, 172,

Index

Male dominance (*cont.*)
Males; and women's thinking, 150, 155, 193n39. *See also* Androcentrism; Gender bias; Masculinity

Males: domestic labor by, 53–54, 75; dominance by, *see* Male dominance; elite white (vs. minority), 73–80 *passim*, 84, *see also* Classism; interests of, in sex difference, 32, 104, 125; and male chauvinism, 15n1, 119n5; and masculine identity, *see* Gender identity; "threat" to, 33n4, 62–68, 79–80, 82, 84, 108, 128, 165 (*see also* Feminist critiques). *See also* Masculinity

"Man" as term, 51n17

"Man-the-hunter." *See* Labor, division of (gender structure)

Marks, Elaine, 28n16, 55n21

Marx, Karl, and Marxism, 26, 33, 42, 47, 89, 109, 146–150 *passim*, 155, 159, 161, 187, 192, 194, 201, 210, 213, 214, 231, 242, 244, 247; post-Marxism, 142–143, 146

Masculinity: abstract, 148–149; authority, value, activity, objectivity equated with, 19, 23, 114, 118, 121, 122, 136, 229; and bias, *see* Gender bias; cultural variations in meaning of, 18, 129; in gender symbolism, 104; and identity, *see* Gender identity; as racial category, 166; of science, 63, 134; separation from mother and, 133. *See also* Males

Mathematics, 43, 48–52, 234

Means, Russell, 165

Medicine. *See* Biology; Neuroendocrinology

Mendelsohn, Everett, 198n2

Merchant, Carolyn, 23n10, 113, 114n3, 123, 124, 128, 129, 136n1, 139n5, 144, 198n2, 225, 233n15, 237n19

Metaphor. *See* Gender symbolism/ totemism

Methodology: experimental, 41, 42, 70, 113, 116, 117, 139, 161n36, 218–222, 224, 228, 239; "feminist," 42, 103; Marxist or Freudian, 33; of physics, as "model," 44, 83; "rule by method," 228–229, 234; selection of problems for study, 22, 25–26, 32, 40, 56, 68, 73, 77, 82, 85, 95, 100, 104, 107, 124, 161, 162, 239; sexism and, 23, 103;

social/androcentric bias in, 24–26, 32, 72–73, 80, 84, 90–94, 105–106, 107, 161, 162. *See also* Objectivity

Meyers, Diana, 171n7

Mill, John Stuart, 109

Millman, Marcia, 22n9, 25n13, 85, 86n3, 89, 90, 106, 161n37

Minden, Shelly, 21n8

Minorities (vs. elite white males). *See* Males

Misogyny, 31, 66, 112, 116, 118, 125–126, 190; Chinese, 145

Money, John, 127nn21–22

Moon, Donald, 35n6, 44n12, 83n1, 211n10

Moraga, Cherrie, 164n1

Morgan, Marabel, 109

Mostowski, Andrzej, 49n14

Motherhood, 22, 112; and "maternal thinking," 184; and mother-child relationship, 132–133; 151–153, 180n21. *See also* Child care; Women

Motivation studies, 21

Moulton, Janice, 51n17

Mulkay, Michael, 198n2

Nagel, Ernest, 211n10

"Naturalists" vs. "intentionalism." *See* Hermeneutics (intentionalism)

Nature: African view of, 169, 171; "domination" of, 16, 76, 168, 171; as machine, 45, 113, 116, 235, 237; uniformity of, *see* Atomism; as woman, 113–118 *passim*, 236, 237

Needham, Joseph, 144–145, 165

Neuroendocrinology, 92, 93, 94–95, 100–101. *See also* Biology

New Science Movement, 217, 219–224, 240–241, 242

Newton, Sir Isaac, and Newtonian physics, 41–48 *passim*, 113, 140, 186, 204, 205, 206, 213, 218, 231, 237, 248

Nietzsche, Friedrich Wilhelm, 27

Nobel Prize(s), 21, 40, 71, 79, 120, 126; woman as winner of, 63

Objectivity, 67; feminist view of, 102, 157, 158, 162; insistence on (as rhetorical device), 67, 117, 135; and "objectivism," criticisms of, 137, 138; of physics, 84, 104, 105; social identity of observer and, 26; of social sciences, 84; vs. subjectivity, 23 (*see also* Gender

Index